The Search for the Absolute

How Magic Became Science

Synthesis Lectures on Engineering, Science, and Technology

Each book in the series is written by a well known expert in the field. Most titles cover subjects such as professional development, education, and study skills, as well as basic introductory undergraduate material and other topics appropriate for a broader and less technical audience. In addition, the series includes several titles written on very specific topics not covered elsewhere in the Synthesis Digital Library.

The Search for the Absolute: How Magic Became Science
Jeffrey H. Williams
March 2020

The Big Picture: The Universe in Five S.T.E.P.S.
John Beaver
January 2020

Relativistic Classical Mechanics and Electrodynamics
Martin Land, Lawrence P. Horwitz
December 2019

The Search for the Absolute: How Magic Became Science
Jeffrey H. Williams

ISBN: 978-3-031-00953-2 print
ISBN: 978-3-031-02081-0 ebook
ISBN: 978-3-031-00153-6 hardcover

DOI 10.1007/978-3-031-02081-0

A Publication in the Springer series
SYNTHESIS LECTURES ON ENGINEERING, SCIENCE, AND TECHNOLOGY
Lecture #5

Series ISSN 2690-0300 Print 2690-0327 Electronic

The Search for the Absolute

How Magic Became Science

Jeffrey H. Williams

Formerly at Bureau International des Poids et Mesures

SYNTHESIS LECTURES ON ENGINEERING, SCIENCE, AND TECHNOLOGY #5

ABSTRACT

History and archaeology tell us that when our far ancestors began to settle in localized groups, they codified their lives and experiences, and formed a collective for mutual support. This proto-civilization would have arisen from each individual's questions about the world, and their attempt to understand themselves and their place in the world. These groups, or tribes, evolved rules of conduct to facilitate communal living, and made a calendar for the group's celebration of harvests, and other events upon which the group was utterly dependent.

This process of social evolution is the origin of religion, and of a magical way of looking at Nature. Eventually, this developing worldview was also the origin of science, which is our investigation of Nature to understand something of what is happening around us, and to use this knowledge to ensure our survival in a violent, indifferent Universe. After all, science and religion seek to answer the same question: Why and how is the natural world the way it is? This book seeks to show how science evolved from religion and magic, in response to a need to understand Nature.

KEYWORDS

origin of science

For Mansel Morris Davies (1913–1995); a man blessed with the gift of friendship.

He not only instructed the author in physical chemistry, but also taught him how to look at the world.

Contents

The world is like the impression left by the telling of a story

(Yoga-Vāsiṣṭha 2.3.11)

Introduction:
Authority and the Collective Memory

History and archaeology tell us that when humans first began to congregate and settle, they codified their lives and experiences, and formed a collective for mutual support. This proto-civilization arose inevitably from each individual's questions about the world, and their attempt to understand themselves, each other, and their place in the world. These groups, or tribes, evolved rules of conduct to facilitate communal living, and they made a calendar for the celebration of important events—events such as planting the crops and when to go hunting and fishing upon which the group was utterly dependent for its survival. These early tribal societies also preserved their songs, their experiences or history, and their stories, fables, wisdom, and beliefs in the memories of the tribe's Shaman or Bard. These collective memories led to myths and legends, which were extravagant and hence memorable, short-hand records of matters such as: invasions, migrations, conquests, dynastic changes, admission and adoption of foreign religious cults, and of social reforms.

This inevitable process of social evolution is also the origin of religion, and of a magical way of looking at Nature; both of which are still with us today. Eventually, this evolutionary process was also the origin of science, which is essentially our investigation of Nature to understand something of what is happening around us, and to use this knowledge to ensure our survival, and the survival of our tribe or extended family in a violent, indifferent Universe. After all, science and religion seek to answer the same two questions: (1) Why and how is the natural world the way it is? (2) How best can we assure our survival? In addition, myths and science fulfill a similar function: they both provide man with a representation of the world, and of the forces that are supposed to govern it. Both myths and magic, together with science, fix the limits of what is possible.

At first, the members of the tribe were easily cowed and controlled by the superstitious fear associated with mythology; such control cannot be generated in a group without myths and marvels. Thus magic, or proto-science, was at the heart of the social organization of early societies. But the essential part of this tribal codification of experiences was the recording of the information necessary for survival; that is, the tribal wisdom. For example, the recording for future generations of the hard-won collective experience that to sustain the life of the tribe, crops are best planted at a certain moment of the Solar Cycle, and harvested a certain number of Lunar Cycles later; that fish are best looked for at a high tide (again the relationship with the Moon), and that large animals are best hunted in early morning when they are rutting at certain times in the Solar Cycle. But how does a non-literate society define those dates, and how did they determine when a particular date was approaching?

The readily observed phases of the Moon formed man's first chronometer, and by hard experience the tribe would have learned the most appropriate time, relative to the phases of the Moon and the Solar Cycle to go hunting and gathering. Later, a combination of a Lunar Calendar and a Solar Calendar told agricultural man when was best for planting and harvesting. Our ancient ancestors (who were probably no less insightful than we are today) would also have noted the similar and hence, perhaps, related time scales of Lunar Cycles and female fertility; hence, they evolved a Moon goddess, rather than a Moon god to represent this fertility and to whom supplications could be addressed, although the desired "responses" would only be forthcoming on a statistical basis, but this was evidently good enough. This tribal knowledge was so important that it needed to be preserved for the survival of future generations. Thus, a calendar based upon an understanding of astronomy and biology was one of the essentials for the survival of the earliest human communities, and, indeed, this remains the case for that shrinking part of humanity that does not live in towns and cities.

This tribal wisdom or language of survival, which originally would have been passed down by the tribe's Shaman or Bard, would have been articulated in the spoken language of the tribe. Other tribes would have learned the same essential things, but would have spoken about those same things in a different vernacular. Thus, as various spoken languages developed, they were incorporating, and permitting, the transmission of the same essential knowledge; that is, how to survive and prosper. This was the beginning of science, which was of necessity supra-tribal, and later became supra-national. Consequently, common to all languages and cultures is a set of observations and facts; what today we call, in its most general form, science, but was in fact astronomy and biology. But back in the distant past, our ancestors would have considered all this magic, or perhaps religion.

The question we shall investigate in this volume is: How did the modern form of the language of science arise from earlier tribal wisdom? Put another way: How did man develop a worldview which allows him to classify and understand all the phenomena and things observed in Nature? The modern, technical language of science is actually very simple as a language, but it has grown very far from the vernacular languages of literature. Yet, the clearly defined language of science, which is taught to students from a young age (although few realize that they are being taught a language different for the vernacular they use with each other) is the nearest thing that we have to a universal or perfect language; that is, a language that can be understood by all men cutting across the confusion and redundancy of vernacular languages.

In these pages, we will see how it was that in attempting to construct a complete set of the observations needed by prehistoric man for his survival, and the survival and stability of his tribe or extended family, we were inevitably led to the development of a system of classification that best facilitated the transmission of this information. The earliest proto-scientists, or natural philosophers, or Shamans, or magicians realized that instead of learning long lists of natural phenomena, and of biological information and astronomical observations that would assist their society to survive the potentially fatal vagaries of Nature (e.g., climate change) it was more logical, and a lot simpler, to arrange the essential facts into different classes (which have today evolved to become different sciences) and then attempt to find a principle of coherence behind all these observed facts. Such a systematization would render the long lists irrelevant, thereby saving everyone's limited and imperfect memories. It also permitted the more insightful natural philosophers, or proto-scientists (and some of these early scientists also practiced magic) to begin making predictions about the working of Nature, thereby creating modern science and technology. But then, science like magic and religion was always interested in everything. It was the epistemological earthquake that was the French Revolution that gave us separate, non-communicating, independent disciplines and schools of thought.

A CALCULUS OF THOUGHT

First, we will explore how and why we record essential information. I am sure that I am not alone in that, when confronted by the complexity of daily life. I find it a great relief to make a list. The preparation of the list allows me to put my thoughts in order by putting them down on paper. I am taking control of some aspects of my life, and instilling order into part of the chaos that surrounds me. This fetish with list-making probably stems from my student days, when my lists of things I had to learn were very long, but by the time I graduated they were considerably shorter and more concise. And in so doing, as we will see. I had not only earned a degree in natural science but I had also trained my memory in the manner of the Catalan mystic, the Blessed Ramón Llull (c.1232–c.1315) and his later follower, the Catholic heretic and *savant* Giordano Bruno (1548–1600)—two key figures in the early part of the search for the universal language of science (see Figure I.1).

Figure I:1. Tree of Science (*Arbor Scientiae*) is one of the most extensive manuscripts of the 13th Century Ramón Llull, written in Rome between 1295 and 1296. It is a version of the author's *Ars magna* written for a general readership (see Chapter 1). It is one of the first attempts in Europe to describe the holistic nature of science, that is, the oneness of Nature, and to try and communicate this idea to a wide readership. As we can see, the work uses a familiar analogy: the organic comparison, in which science is represented by a tree with roots, trunk, branches, leaves, and fruits. The roots represent the basic principles of science; the trunk is the structure; the branches, the genres; the leaves, the species; and the fruits are the details. This vegetal allegory shows the influence of Aristotle. This image will serve as a metaphor for this work, and is taken from https://en.wikipedia.org/wiki/Tree_of_Science_(Ramon_Llull).

However, list making as a means of trying to order the overwhelming quantity of information we all come upon in our lives is not a new concept. At the dawn of literature, Homer presents us with the two possible ways in which information, or data, may be presented and stored for future reference. That is, either as a simple long list, or as a closed-system which contains all knowledge in microcosm and which shows us, the observers, how all things are interconnected in miniature—into which one must know how and where to look to find what it is one is seeking to understand, or to know. A list brings order, and through its use we can (at least) try to understand, influence, and perhaps control the world around us. We are able to exclude things; creating a list is a means of making choices. One might imagine that a list seeking to represent a complex set of information such as an entire discipline of science would produce a near infinity of possibilities and so be useless, but lists actually bring their own rules and orthodoxy. We will look at how it was that we moved from merely making and trying to memorize long lists of observations to a rationalization of such lists in terms of an underlying principle: the move from the qualitative to the quantitative. The *I Ching* of Ancient China is a good example of this evolutionary move from a magical worldview toward a truly empirical, science worldview.

The earliest examples of a scientific or philosophical language, those from before the 17th century, were not quantitative. The language used by the proto-scientists when they communicated among themselves was purely qualitative. The early experimenters, or alchemists, were not overly keen to discuss in too much detail what it was they were doing and why they were doing it. Consequently, early manuscripts read more like a mystery story or a philosophical explanation than a description of an experiment and the resulting observation of the consequences of the experiment. But then, these proto-scientists were living a dangerous life; the Church would have condemned them and burned them if their actions were clearly described. There was safety in obfuscation and cloudy philosophical concepts. However, the baleful influence of the Church, and its own inability to effect any change in Nature (miracles) did eventually decrease in importance in society; the purely statistical success rates of prayer were finally deemed not to be good enough and eventually a quantitative language of science would be invented. Such a quantitative scientific language was naturally capable of extension, leading to explanation and prediction, able to support international communication and commerce. It was a new *lingua franca*, but a language devoid of metaphor and multiple, confusing meanings. This was, of course, not a new idea. The idea that there once existed a perfect language, which was spoken by all mankind, has occupied the minds of *savants*, mystics, Neo-Platonists, natural philosophers, and theologians for well over two millennia. This language was perfect in that it expressed without ambiguity the essence of all things and ideas—the quiddity of all things. It was a language in which there was only one possible way of describing, e.g., an animal, a natural phenomenon, or an explanation of why something happened. It was also accepted that if this perfect "language of Eden" could be recovered, men would again be able to comprehend each other fully and comprehend the functioning of Nature and thus the meaning of existence. Men would be able to abolish discord and strife, and return to a Golden Age.

Even today, there are still physicists seeking to discover the perfect universal language in the form of the Theory of Everything (Chapter 11), although the majority of the physicists and mathematicians researching this project do not appreciate the immensely old tradition within which they are laboring. The discovery of the Higgs' boson, and of gravitational waves, are only the latest steps in man's quest for the absolute, for the essence of the natural world, and for a single, unambiguous Theory of Everything. The development of modern science may, in large part, be considered as stages in this investigation; an attempt to understand the "make of all things." Certainly, the creation of the system of quantities and units, which today we call the International System of Units (SI from its French official name, *Système International d'Unités*), during the French Revolution was pivotal in allowing man to finally abandon magic and mysticism, and the memorizing of long, tedious, incomplete lists of properties and observations in his investigation of Nature, and to adopt a coherent, scientific worldview.

CHAPTER   1

In the Beginning Was the List

There must be a beginning in any great matter but the continuing unto the end, until it be thoroughly finished yields the true glory.

Sir Francis Drake (c.1540–1596)

Some may be surprised to read that there is a link between magic and modern science; that modern science evolved out of magic. Indeed, one could go further and state that modern physics and chemistry would not exist had it not been for the ideas and "experiments" of the Neo-Platonists of the early-Christian world. The problem is that there is a spiritual aspect to Neo-Platonism; there is more metaphysics than physics in Neo-Platonism, and so many contemporary physical scientists would be aghast at a suggestion of the metaphysical origins of their subject.

But you do not have to go too far into the quantum mechanical explanation of spin-entanglement and the mixing of quantum states, that is, the generation of qubits of quantum information, before you realize that what you are dealing with is more philosophical than physical. Today's laser physicists attempting to teleport quantum information[1] from a laboratory on one continent to a laboratory on another continent are having to revaluate what is actually meant by a "measurement," to fully comprehend their results. These modern physicists are undergoing the same self-analysis that the Taoists recommended to all natural philosophers, and which the alchemists sought in their explorations of Nature (see Figure 1.1), although few contemporary physicists have ever thought about their sophisticated experiments in this way.

History tells us that before there was science, and its most useful offshoot, technology, there was magic, and a magical way of looking at the world. In the evolution of a culture, the scientific worldview is always a late development. In the evolution of our culture, the 17th century supposedly marked the period when astrology, the burning of witches, and folk-magic yielded to Isaac Newton's rationalism, and the Laws of Nature were established as observation and experience explained

---

[1] A pure qubit state is a coherent superposition of the states' wave functions. This means that a single qubit can be described by a linear combination of the wave functions $|0\rangle$ and $|1: |\psi\rangle = \alpha|0\rangle + \beta|1\rangle$, where α and β are probability amplitudes. When we measure this qubit in the standard basis, according to the Born rule, the probability of outcome $|0\rangle$ with value 0 is $|\alpha|^2$ and the probability of outcome $|1\rangle$ with value 1 is $|\beta|^2$. Because the absolute squares of the amplitudes equate to probabilities, it follows that α and β must be constrained by the equation $|\alpha|^2 + |\beta|^2 = 1$. Note that a qubit in this superposition state does not have a value between 0 and 1; rather, when measured, the qubit has a probability $|\alpha|^2$ of the value 0 and a probability $|\beta|^2$ of the value 1. In other words, superposition means that there is no way, even in principle, to tell which of the two possible states forming the superposition state actually pertains. [1]

by reason. Yet before there was reproducible and reliable science, there was unreliable or "chancy" science. Even in the Renaissance, scientific work (what would then have been termed an "exploration of natural philosophy") was a hit-and-miss affair, as few *savants* noted down the details of what it was they had done in their experiments. And as quantities of substances were not measured consistently, or measured at all (quantities of chemicals and materials could not even be defined precisely, as units of measurement were entirely parochial), and the materials used were of varying degrees of purity, experimental science was the affair of each individual practitioner. Consequently, at this time both science and magic were acceptable and interchangeable ways of interpreting Nature, as neither one nor the other was infallible or even reproducible; both appeared to work only on a statistical basis. And so the experimentalists would have believed they that had not been, for example, in the "right frame of mind," or the stars were in the wrong alignment on the day their experiment did not yield the result it was supposed to yield.

Indeed, the scientist or *savant* of that time dabbled in both natural philosophy and the occult. Isaac Newton (1643–1727) was himself something of a magus or, at least, a Neo-Platonist. At the tercentenary of Newton's birth, John Maynard Keynes (Newton, *The Man*, lecture given as part of the Royal Society tercentenary celebration of the birth of Newton) described him as the last of the magicians, "*Newton was not the first of the age of reason. He was the last of the magicians, the last of the Babylonians and Sumerians, the last great mind which looked out on the visible and intellectual world with the same eyes as those who began to build our intellectual inheritance rather less than 10,000 years ago.*" Isaac Newton was a man with an immense, insatiable curiosity, for whom nothing could, or should, be taken at face value; he was above all a man interested in exploring everything, and spent much of his time in his laboratory doing experiments. Newton examined everything, including several stomach-turning experiments on his own eyes. Today we may consider ourselves as rational, coldly logical, non-superstitious, scientific beings with several degrees of separation from those who believe in magic and superstition. In the time of Newton, however, there were fewer degrees of separation between such individuals—if any at all. But by the end of Newton's life, the European Enlightenment was underway and the triumph of science over hermeticism and religion was more or less assumed. But as it turned out, there was a huge revival of occultism and the hermetic arts in the late-19th century (Chapter 16).

Before beginning our examination of the origins of science, let us consider the purpose of science, and the object of scientific investigation, that is, the understanding of Nature. Perhaps the most famous example of a magician in literature is Faust. In the earliest sections of Johann Wolfgang von Goethe's great poem (begun c.1772), the magician and his tempter correspond broadly to traditional mediaeval figure (the disillusioned old man manipulated by the Devil) and the plot gives a somewhat traditional, Christian version of the concepts of salvation and perdition. However, by the time that Goethe finished his poem (1831), Faust had evolved. He is no longer a magician excited and led astray by his desire for forbidden knowledge; Faust has been transformed into the

Romantic figure of Everyman. Faust has become a seeker for oneness with all Nature. Between 1770 and 1830, our civilization had moved from the Classical world to the Romantic world, and today, we are all still lost in the late-Romantic world. The later, holistic Faust has abandoned a Manichaean dualism of good and evil, for a mystical sense of the unity of all things. This Romantic Faust would likely have been an early recruit to the politics of environmentalism. Faust no longer embodied heterodox magic, but acceded to a knowledge of the interconnectedness of all natural phenomena (which is the way modern science views Nature, see Chapter 9). In this way the character of Faust follows the evolution of magical and scientific thinking; first there was magic, and then there was physics; first there was the sorcerer, and then there was the physicist.

Figure 1.1: *The Alchemist*, a painting by Joseph Wright of Derby (1734–1797). An image of a man searching, both experimentally and mystically, to understand himself and his place in Nature. Image from: https://en.wikipedia.org/wiki/Joseph_Wright_of_Derby#/media/File:Joseph_Wright_of_Derby_The_Alchemist.jpg.

The magic that predates, and inevitably leads to, science is a force which follows processes and events that are inherent to consciousness, and is something implicitly connected to constructive and imaginative thought, therefore to the whole enterprise of artistic and scientific creation. Our imaginations, our dreams, our ability to use our consciousness to imagine and to describe, and then to transfer theories and fantasies, are inherently bound up with our facilities for reasoning. And they are essential for making that great leap from observing and explaining a known phenomenon, to going beyond into the realms of prediction. The fabulous and the fantastic are all around us. Often, the more we examine a phenomenon that was once deemed to have been fully comprehended, the more we may learn, which allows us to speculate and then, perhaps, realize that the fantastic is not entirely separate from the natural.

In the world before the 18th-century Enlightenment, magic and fantasy were inextricably linked with man's attempt at understanding the forces of Nature, but nonetheless this pre-scientific worldview led to considerable advances in technology, and to many useful discoveries in engineering, metallurgy, chemistry, and pharmaceuticals. When thinking about the world in a magical sense, one considers the phenomena of Nature to have arisen through the agency of certain secret forces of Nature; forces which may reside uniquely in certain objects (for example, the loadstone of the flying island of Laputa in Jonathan Swift's *Gulliver's Travels* of 1726, which would have relied upon repulsive magnetic fields (see Figure 12.1), which were not fully explained until the late 19th century; see Chapter 12) or with certain inspired individuals, for example, Giordano Bruno, Leonardo Da Vinci, or Philippus Aureolus Theophrastus Bombastus von Hohenheim (Parcelsus). Such magical thinking structures the processes of imagination; and imagining something can, and sometimes does, precede the fact, or the act of discovery. This apparent breakdown in causality is something that we would today term intuition or instinct, and is something that a great many people experience without thinking about it. Indeed, you do not have to investigate modern quantum mechanics and the quantum-view of nature very deeply before concepts such as causality and non-causality, of cause and effect, become much more confused than one would have ever supposed. Such concepts become, essentially, metaphysical. [2]

1.1   THE LIST AS THE ORIGIN OF SCIENCE

But let us return to the very dawn of science. How was it that the earliest observers of Nature noted what it was they had observed? Not having a theory to explain their observations, it was likely that they merely made a list of what they had seen. Thus, the earliest stage of the evolution of science was the creation of long lists of things and events. Indeed, these long lists of observations and results would likely have been compiled of symbols (as there was no standard nomenclature of chemicals or phenomena), numbers, and words in a vernacular (see Figure 1.2). Of course, the natural philosopher may also have written his notes in a cypher; for example, the Angel Language of the Elizabethan alchemist John Dee.

List making is a means of attempting to control and understand the complexity of life and of the world around us, and of trying to order the seemingly incomprehensible quantities of information we all come upon in our lives—not a new concept. It is simply a way of recording information or data. But it is not the only way. At the dawn of literature, Homer presents us with the two possible ways in which information, or data, could be presented and stored for future reference. Either as a simple but long list, or as a closed system which contains all knowledge in microcosm and shows us, the observers, how all things are interconnected, but into which one must know how and where to look to find what it is one is seeking to know, or to remember; what today we would term a database.

Figure 1.2: A table of alchemical symbols from Basil Valentine's *Last Will and Testament* of 1670. (Image from: https://en.wikipedia.org/wiki/Alchemical_symbol#/media/File:Alchemytable.jpg). Basil Valentine, or Basilius Valentinus, was supposedly a 15th-century alchemist and Canon of the Benedictine Priory of Saint Peter, Erfurt, Germany. But this name could also have been an alias for a number of German chemists/alchemists. During the 18th century, it was suggested that the author of the works attributed to Basil Valentine was Johann Thölde, a salt manufacturer in Germany (1565–1624). Whoever he was, Basil Valentine had considerable ability and ingenuity as an experimental chemist. He showed that ammonia gas could be obtained by the action of alkali on sal-ammoniac (ammonium chloride), described the production of hydrochloric acid by acidifying brine (sodium chloride), and created oil of vitriol (concentrated sulfuric acid). Thus did modern chemistry grow out of alchemy; particularly, when the spiritual aspect of the alchemist's craft lost its precedence.

In Book 2 of Homer's *Iliad*, we encounter what is called *The Catalogue of Ships*. This catalogue is a long list of the various Greek forces who came together to attack Troy, and Homer uses it as a means of structuring information which the reader will need later in the story. However, in Book 18 of the *Iliad*, Homer presents us with a very different manner of presenting and storing information; he gives us a description of the shield the lame-god Hephaestus has made for swift-footed Achilles after the death of Patroclus. This shield was a microcosm within which all of Nature was to be found; the Sun, the Moon, the 12 Houses of the Zodiac, and the images of the world of men through the changing seasons of their year, all bound by the mighty river of Ocean running around the rim of the shield. This shield allowed the viewer, provided he knew where and how to look, to find whatever piece of information about the lives of men, or of astronomy and agriculture, he needed, or all the possible ways that these pieces of information could be coupled together. This shield was undoubtedly intended to promote the semi-divine nature of Achilles, and to protect him in battle. But even Hephaestus could not protect lion-hearted Achilles from his fate, not even with a database of all human knowledge on his arm.

A list brings order, and through it we can understand and try to control the world around us. We are able to exclude things; creating a list is a means of making choices. One might imagine that a list seeking to represent a complex set of information such as the ephemerides of the Sun and Moon, or all the possible ways of reacting a number of the basic chemicals used in alchemy, would produce a near infinity of possibilities and so be useless, but lists actually bring rules and orthodoxy. Homer did not list all the petty kings of Bronze Age Greece, he only listed those relevant to the story he was about to relate of the Trojan War. The author had thought about these petty rulers and then placed them in the wider society and culture of the Ancient World. In a similar way, Homer had thought about Achilles' shield, and designed it to represent the world in miniature. In other words, Homer had thought about the essential elements of Nature and how they interact, and was presenting this summary of the essential points to future generations as a database.

Another example of this design, by an author of a detailed model of the world, which was then put on paper for future users is that greatest of all poems, the *Divine Comedy* by Dante Alighieri (1265–1321). In the 14,233 lines of this masterpiece, Dante gives us a complete representation of the medieval worldview; a concise summary of the Aristotelian-Thomist cosmogony of the late 13th century. The poem is a portable, readable summary of everything that a Christian needed to know to achieve salvation, and to understand the natural world in which he finds himself.

But a list can also be a frightening thing. Our imperfect memories will always tell us that we have forgotten something, and that this something is hugely important. And the more we try desperately to remember that important something, the more it slips from our mental grasp. Lists can be troubling, even subversive. Our lives are limited; death is a particularly discouraging limit. This is why we all like subjects of investigation that have no limit, and therefore no end, for example, history, science, and philosophy. Long lists are a way of escaping from our thought about death. Even if a list makes us anxious about things we cannot remember we like it because we do not want to die. [3]

Making lists, or pictorial or text-based summaries of a field of knowledge, may impose order, but what is really required to effectively use the contents of any long list is a means of manipulating all the possible entries in the lists of all the various categories of all objects and all ideas. This could rapidly become a vast quantity of data. And what of the nature of science? Anyone who has ever attempted to study science knows that there is an awful lot of memory work to be done. Although all science students are told that the vast edifice of science rests on a few basic axioms and theories, it is unfortunately true that before one can get to study these foundation stones of science you have to spend years memorizing seemingly incomprehensible amounts of information, data, rules, and exceptions to the rules. Of course, the professors will tell the aspiring scientist that he or she must first be grounded in the factual matter of the subject before they can make an attempt at comprehending, and perhaps applying or even extending, the basic axioms. But then professors and teachers have always said that—in every discipline. What, of course, should be done is to explain and teach the axioms to the brightest of the eager students. Then the mountains of facts and generalities could be derived by the students themselves. But, alas, that is not the way the teaching of science, or of any other subject, has evolved.

This is the heart of the problem to be examined in this book; how do we construct a simple language of few words and few rules, and use this language to describe all the phenomena seen in Nature? How do we take endless lists of observations and facts and reduce, purify, and concentrate them, as the alchemists would have said, to a handful of base units that can be combined in various ways to describe everything we see around us, and everything we will ever see?

1.2   RAMÓN LLULL

Perhaps the first person to attempt a method of systematizing and classifying information for the purpose of facilitating the compilation and generation of derived information systems was the Catalan mystic, the Blessed Ramón Llull, or Raymond Lull (c.1232–c.1315). He designed a teaching aid that was also a means of deriving or generating information from lists, which he called his *Ars generalis ultima*, or *Ars magna* (The Ultimate General Art, or Great Art), which appeared in 1305. This was an attempt to combine, or manipulate, the contents of long lists, in particular, to generate the theological and philosophical attributes of the Divine, selected from a number of lists of those attributes.

Ramón Llull intended this technique of combining and manipulating lists of theological and philosophical information to be a debating tool for converting Muslims and Jews to Christianity using logic and reason. Through his detailed analytical efforts, Llull built a theological reference by which a reader, or proselytiser, could enter into a mediaeval calculating machine (it was called Llull's "thinking machine") an argument or question which had been put to them about the Christian faith. The reader would then turn to the appropriate index and page in the Llullian calculator to find the correct answer to the question posed by the potential convert.

The radical innovation of Llull was the introduction of logic into list making. In particular, the construction and use of a "thinking machine" made of inscribed metal discs to combine elements of thought; for example, elements of language. With the help of connected geometrical figures, following a precisely defined framework of rules, Llull tried to produce all the possible statements of which the human mind could conceive on certain subjects. These declarations or statements were represented by a series of signs, or sequences, of letters derived from his proto-computer, or "thinking machine." We will now briefly consider the hardware and software (see Table 1.1) of Llull's "thinking machine."

Table 1.1: The alphabet of Llull's thinking machine. This is the software of the device built by Ramón Llull to explore the logic of list making. It contains the theological and philosophical attributes of the Divine. The letters in the first column of the table (contain the "computer programme" of Llull's device) correspond to the outer circle of the hardware, that is, to the engraved metal discs shown in Figures 1.3 and 1.4

	Figure A (given in Figure 1.4)	Figure T (given in Figure 1.5)	Questions and Rules	Subjects	Virtues	Vices
B	Goodness	Difference	Whether?	God	Justice	Avarice
C	Greatness	Concordance	What?	Angel	Prudence	Gluttony
D	Eternity	Contrariety	Of what?	Heaven	Fortitude	Lust
E	Power	Beginning	Why?	Man	Temperance	Pride
F	Wisdom	Middle	How much?	Imaginative	Faith	Accidie (apathy)
G	Will	End	Of what kind?	Sensitive	Hope	Envy
H	Virtue	Majority	When?	Vegetative	Charity	Ire
I	Truth	Equality	Where?	Elementative	Patience	Lying
K	Glory	Minority	How and with what?	Instumentative	Pity	Inconstancy

The software of Llull's device, given in Table 1.1, is essentially an alphabetic list giving the meaning of nine letters, in which Llull (the programmer) says "**B** signifies goodness, difference, whether?, God, justice, and avarice. **C** signifies...," and so on (there is no J in the mediaeval Latin alphabet). The components of the first column of Table 1.1 are set out in Llull's *Figure A* (that is, Figure 1.3 here). The letters don't represent variables, but constants. Here they're connected by lines to show that in the Divine these attributes are mutually convertible. That is to say that God's goodness is great, God's greatness is good, etc. This, in turn, was one of Llull's definitions of God, because in the created world people's goodness is not always great, nor their greatness particularly good, etc. Such a system of vertices connected by lines is what mathematicians term a graph. This might seem to be of purely anecdotal interest, but as we shall see shortly, the relational nature of Llull's system is fundamental to the idea of an *Ars combinatoria*.

The components of the second column in Table 1.1 are set out in Llull's *Figure T* (that is Figure 1.4 here). Here we have a series of geometrical principles related among themselves in three groups of three, hence the triangular links. The first triangle defines: difference, concordance, and contrariety; the second defines beginning, middle, and end; and the third triangle defines majority, equality, and minority. The concentric circles between the triangles and the outer letters show the areas in which these relations can be applied. For example, with the concept of difference, notice how it can be applied to sensual and sensual, sensual and intellectual, etc. "Sensual" here means perceivable by the senses, and Llull explains in the *Ars brevis* that: *"There is a difference between sensual and sensual, as for instance between a stone and a tree. There is also a difference between the sensual and the intellectual, as for instance between body and soul. And there is furthermore a difference between intellectual and intellectual, as between soul and God."*

This hardware consists of three inscribed metal disks fixed on a single axis on which they can be rotated independently. The disks contain a limited number of letters—a special lullistic alphabet. When the circles are turned, step-by-step, all possible combinations of these letters are produced. The metal-circle called the *Prima Figura* (Figure 1.3) gives the primary attributes. The next strictly defined table of words can be produced on the next circle, Secunda Figura (Figure 1.4), where we find categories and relations of thinking.

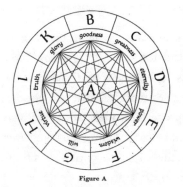

Figure 1.3: A list of the attributes of God (see second column in Table 1.1). This is Llull's *Figure A* or *Prima Figura*. Image from: http://www.ramonllull.net/sw_principal/l_br/home.php.

Figure 1.4: A list of the categories and relations of thought (see third column in Table 1.1). This is Llull's *Figure T* or *Secunda Figura*. Image from: http://www.ramonllull.net/sw_principal/l_br/home.php.

Ramón Llull's thinking machine allows all the words (attributes of the Divine) in the outer circle to be combined in different ways by turning the circles, relative to each other in a stepwise manner. It is therefore possible to connect every word with every other word placed in a position of a table—depending only on the construction of the individual tables. Ramón Llull created numerous devices for this manipulation, or combining of the contents of lists. One method is now called the Llullian Circle, which consisted of two or more solid circular discs, a disc of smaller diameter free to rotate inside a larger annular disc; both these independently rotating discs were inscribed with alphabetical letters or symbols that referred to, for example, components of lists of attributes of the Divine. A number of terms or symbols relating to those terms were laid around the circumference of the circle. The discs could be rotated individually, like a circular slide rule to generate an enormous number of combinations of ideas. Thus, the innermost disc could be inscribed with what Llull termed the "absolute characteristics" of God: goodness, eternity, power, volition, virtue, truth, glory, or wisdom, and it could be rotated to be next to attributes on the next outer disc, which was inscribed with "relative characteristics" such as greatness or extent or purpose. Llull conceived this combinatorial manner of generating ideas as a perfect logical language, which could be used to convert non-Christians to Christianity. The language was to be universal; it was to be articulated at the level of expression in rational mathematics, and its content was intended to consist of a network of universal ideas held by all people. Llull based this device on the notion that there were only a limited number of basic, undeniable truths in any field of knowledge, and that we could understand everything about these fields of knowledge by studying combinations of these elemental or fundamental truths.

1.3   DETAILS OF THE *ARS COMBINATORIA*

Given a number of different elements n, the totality of the possible arrangements that can be made from them, in any order is expressed by their factorial $n!$, calculated as $1.2.3. .... n$. This is the method for calculating the possible anagrams of a word of n letters (as in the art of *Temurah* in the Kabbalah). As n increases, the number of possible arrangements rises more rapidly: the possible arrangements for 26 letters of the alphabet would be a vast, incomprehensible number of combinations. If the strings of combinations admit repetitions, then the number of combinations rises even further. Consider the situation of four people. We want to arrange these four as couples on a train, where the seats are in rows that are two across; the order is relevant because we wish to know who will sit at the window. This is a problem of permutations; that is, of arranging n elements, taken t at a time, taking the order into account. The formula for finding all the possible permutations is $n!/(n-t)!$ Suppose, however, that the order is irrelevant. This is a problem of combinations, and we solve it with: $n!/t!(n-t)!$

This is an expression-system (represented both by the symbols and by the syntactic rules establishing how n elements can be arranged *t* at a time—and where *t* may coincide with *n*), so that the arrangement of the expression items can automatically reveal possible content-systems. In order to let this logic of combination or permutation work to its fullest extent, however, there should be no restrictions limiting the number of possible content-systems (or worlds) we can conceive of. As soon as we maintain that certain universes are not possible in respect of what is given in our own past experience, or that they do not correspond to what we hold to be the laws of reason, we are, at this point, invoking external criteria not only to discriminate the results of the *ars combinatoria*, but also to introduce restrictions within the art itself.

This combinatorial method of Llull was an early attempt at using logic to retrieve knowledge, data, or information from a list; that is, to use mechanical means to generate concepts, which avoided the tedious necessity of memorizing vast numbers of combinations of ideas and thoughts. Of course, most of the combinations of ideas so generated would be redundant (in the definition of the Divine, what exactly is the difference between "glorious eternity" or "eternal glory"). Llull hoped to show that Christian doctrines and dogma could be arrived at from any starting point, from a fixed set of preliminary, arbitrary ideas. Llull knew that all believers in the monotheistic religions would agree with the absolute attributes of God, giving him a firm platform from which to argue that the Christian interpretation of the Divine was the most apposite interpretation, and so his listeners would be convinced of his logical vision and convert to Christianity. Whether or not Llull's system of logical persuasion worked for its intended purpose we do not know; history tells us he was nearly killed at the age of 83 attempting to convert Muslims in North Africa.

One can ask, what exactly is Ramón Llull's place in the history of computers and computing? Llull is one of the first people who tried to make logical deductions in a mechanical, rather than a mental way; that is, based on the contents of his imperfect memory. His method was an early attempt to use logical means to generate and retrieve information. He demonstrated in an elementary, but nevertheless workable way that human thought can be described and even imitated by a mechanical device. This was a small step toward the thinking machine of the contemporary world.

The ideas of Ramón Llull about the systematisation or ordering, and the generation and retrieval of information and knowledge were developed further in a more esoteric manner by Giordano Bruno in the 16th century (but unfortunately for Bruno this was one of his undertakings that led him to be burned as a heretic by the Catholic Church in Rome, February 17, 1600), and subsequently by the great German *savant* Gottfried Leibniz (1646–1716) in the late 17th century for investigations into the philosophy of science. Leibniz gave Llull's idea the name *ars combinatoria* (the art of combinations). Many consider Llull's ideas on systems of logic, and their use in constructing new combinations of ideas as the beginning of the study of information science and semiotics. Although the combinatorial calculus of Ramón Llull was an extraordinary creation for the Middle Ages, giving us the earliest machine language and a perfect means of creating unbreakable encryption codes, Llull's reputation was not always as high as it is today. The satirist Jonathan Swift ridiculed such Llullian devices in the third part of *Gulliver's Travels*, where on the Island of Laputa Lemuel Gulliver is shown several large folio volumes of broken sentences generated by the *ars combinatoria* which he is told ".. [will] *piece together; and out of those rich Materials to give the World a compleat Body of all Arts and Sciences...*" Likewise, in François Rabelais' exuberant, witty, and satirical comment on his contemporary world, *The Life of Gargantua and of Pantagruel* the combinatorial arts of Llull are disparaged; Gargantua advices his son Pantagruel to master astronomy "*but dismiss divinatory astrology and the art of Lullius as fraud and vanity.*"

1.4   FURTHER READING

1   Two excellent, readable accounts of the difference between the real and the quantum worlds are: *The Nature of the Physical World* (1947); Sir Arthur Stanley Eddington; London, Dent & Sons Ltd., and *QED: The Strange Theory of Light and Matter* (1985); Richard P. Feynman; Princeton, Princeton University Press.

2   The complex and close relationship between magic and science is touched upon in many places in *Stranger Magic: Charmed States and the Arabian Nights* (2011); Marina Warner; London, Chatto and Windus.

3   Everything you have ever wanted to know about lists and list-making: *The Infinity of Lists* (2012); Umberto Eco; Maclehose Press (an imprint of Quercus).

4   What is the definition of magic, and of a magical way of looking at the world? *A General Theory of Magic* (1972); Marcel Mauss; London, Routledge; originally published in French in 1902.

CHAPTER   2

The Origins of the Language of Power that Is Science

And God said, Let there be light: and there was light.

Genesis 1: 3-4

Today, if you talk to a scientist about what it is that he/she believes science represents, you will likely be told that science is the only real source of truth; that the ideas of science are not culturally specific. That these scientific truths are as comprehensible for an American, as they are for a Russian or a Chinese *savant*. Of course, the American, Russian, and Chinese *savants* must be able to read their respective languages, and it is unlikely that anyone of them would be able to read the other two languages sufficiently well to comprehend what it was that the texts were describing. But the idea that science has an international, non-cultural character, not dependent upon particular elements of vocabulary, or particular rules of grammar, is certainly accepted by most scientists.

About a thousand years ago it would still just have been possible for the majority of scholars and *savants* to have been able to discuss something through the medium of a true international language such as Latin, Arabic, or Greek. But such direct communication has not been possible since the end of the Middle Ages and was only ever possible in some parts of the European and Mediterranean worlds. Ancient Chinese scholars, like their Greek contemporaries, believed that their language was the only appropriate medium of communication, so they did not concern themselves with the languages of barbarians.

So, what exactly is the origin of this confusing association of science with a medium of universal communication? Indeed, can science be a universal medium of communication? This question is of particular interest, given the marked inability of the majority of scientists to explain to non-scientists, and even to scientists in areas of specialzation not their own, what it is they do and why it is that they do what they do. Therefore, what does it mean to say that science is a form of universal language, a culturally independent means of communication?

Let us look at "the babble after Babel;" that is, the myth of the origin of the multitude of languages that we know today. The idea of an age when all men could converse with each other and could converse directly with their Creator, who had taught them this *Ur*-language, is a very widespread myth.[2] It is a myth about a Golden Age of innocence; a myth known to every culture—it is an idea deeply rooted in the Jungian collective consciousness. But for all its mythic qualities, the search for this ideal language, which according to monotheist beliefs had also been spoken by God to bring the Cosmos and humanity into existence (see the quote at the head of this chapter), obsessed European philosophers and *savants* right up until the Age of Enlightenment, when the Indo-European theory of the origin of modern languages was developed, and accepted.

In *Genesis* 2:16-17, we are told that the Creator God spoke to man for the first time; telling our ancestor, Adam that everything in the earthly Paradise was his. God commanded him, however, not to eat of the fruit of the tree of the knowledge of good and evil. We are not told in what language God spoke to Adam. Modern Biblical tradition has imagined it as a sort of language of interior illumination, rather than a communication of words, or of thunderclaps and lightning.

After this command, we read that God "*formed every beast of the field, and every fowl of the air; and brought them unto Adam to see what he would call them.*" Here we have a motif also common to most other religions and mythologies; that of the name-giver, the creator of language (*nomothete*). Figure 2.1 gives a late-mediaeval image of that naming process. Yet it is not at all clear on what basis Adam actually chose the names he gave to the animals. In the Latin Vulgate Bible, we are told that Adam called the various animals "*nominibus suis*" or "*by their own name*s;" and in the King James Bible we have "*Whatsoever Adam called every living creature, that was the name thereof.*" But were the names given by Adam, the names by which each animal ought to have been known, or were they simply arbitrary names given by Adam? That is, did the given Adamic name refer to some fundamental or intrinsic property or characteristic of the animal, or was it purely a matter of what Adam was thinking at that moment.

In *Genesis* 2:23 Adam speaks to his female companion, one again assumes that they are using the Divine Language of Creation as their means of communication, "*This is now bone of my bones, and flesh of my flesh: she shall be called woman...*" Eventually, Adam calls this female companion Eve (which means life as she is to be the mother of humanity) so we see that Adam's choice of name for those things he was charged by God to name was etymologically correct and not arbitrary. This would be a reasonable deduction given that the language in question was itself the language of Creation.

The linguistic theme is taken up again in *Genesis* 11:1. We are told that after the Flood and the repopulation of the earth by Noah's descendants, "*the whole earth was of one language, and of one speech.*" Yet, men in their vanity and arrogance conceived a desire to rival God, and thus erect a tower that would reach up to the heavens. To punish human pride, and to put a stop to the construction of their tower we are told that God devised a plan: "*Go to, let us go down, and there confound their language, that they may not understand one another's speech... Therefore is the name of it called Babel (as*

---

[2] In monotheist religions, the myth is given in the various sacred books. In Hindu culture, it is presented in the *Mahabharata*.

represented in Figure 2.2*); because the Lord did there confound the language of all the earth: and from thence did the Lord scatter them abroad upon the face of all the earth*" (*Genesis* 11:7, 9).

Figure 2.1: Image from the late-Byzantine, 14th-century Orthodox Holy Monastery of Saint Nicholas of Anapafsas, Greece. Adam, in his naked innocence, is seen naming the animals as they pass before him. The dragon-like creature, next to the lion(?) appears to be a remnant of a mediaeval bestiary. Image from: https://commons.wikimedia.org/wiki/File:Adam_naming_animals_-_Moni_Ayou_Niko-laou_(Meteora).jpg.

2.1   A LESS MYTHIC INTERPRETATION OF THE BABBLE AFTER BABEL

Stories accounting for the multiplicity of human languages appear in nearly all mythologies and theogonies. But it is a major leap from knowing that many languages exist to deciding that this multiplicity is a fault or punishment that could be healed by a search for the imagined perfect original language. Indeed, how would you know you had discovered the *Ur*-language, the language of Eden?

For Ancient Greek philosophers, Greek was the language of thinking and ratiocination. This was not a claim that the Greek language was a primary language: it was simply a case of the iden-tification of thought with its natural vehicle. About the speech of barbarians or non-Greeks, the Greeks knew little; hence, little was known about what it would be like to think in the language of barbarians. The Greeks admitted that the Egyptians and the Babylonians possessed wisdom, only because someone (Herodotus) had explained this to them in Greek.

As Greek civilization expanded, the status of Greek as a language also evolved. In the period following the conquests of Alexander the Great (356–323 BCE), a common, universal form of

Greek spread rapidly—the *koine*. This was the language of Polybius, Strabo, Plutarch, Aristotle, and of the Eastern Roman Empire; it was the language taught in the schools of grammar. Gradually it became the official language of the Mediterranean world, and of the East of Alexander's conquests. Spoken by patricians and *savants*, Greek survived under Roman domination becoming the language of commerce and trade, of diplomacy, and of scientific and philosophical debate. It was finally the language in which the first Christian texts were transmitted (Septuagint translation of the Jewish Bible in the 3rd century BCE, and the Gospels in the first centuries AD), and it was the language of the early-Fathers of the Church. A civilization with an international language does not need to worry about the multiplicity of tongues. Nevertheless, such a civilization can, and did, worry about the rightness of its own tongue.

While the Greek *koine* continued to dominate the intellectual life of the Mediterranean world, Latin was becoming the language of the administration of the empire, and thus the universal language for those parts of Europe conquered by the Roman legions. Once again, a civilization with a common language is not troubled by the multiplicity of vulgar tongues. Learned Roman patricians would discourse in Greek, but the rest of the Latin-speaking world needed translators.[3]

Despite this Mediterranean civilization, by the 2nd century AD savants began to study languages other than Latin and Greek, finding that human experience and wisdom could be expressed just as well in other languages. The Greco-Roman world was changing; new religions and beliefs were spreading from the East. Obscure revelations appeared—ome were attributed to Persian *magi*, others to an Egyptian divinity called Thoth-Hermes, to Chaldean oracles, and to Pythagorean and Orphic traditions which, though born in early-Greek civilization, had been buried by rationalist Greek philosophy. Today, we term these mystical beliefs Hermeticism, the product of the mythic Hermes Trismegistus (see Figure 6.1). The classical rationalism, elaborated and re-elaborated over centuries, began to show signs of age. With this loss of rationalism, the established religions entered a period of crisis. The Imperial Pagan religion had become a purely formal affair of the law courts; a simple expression of loyalty to the state. Each conquered people had been allowed to keep its own gods. And these new gods were, as in all conquering empires, accommodated to the Latin pantheon; no one bothering about contradictions, synonyms, or homonyms.

A result of this widespread syncretism was the creation of modern monotheism, with its belief in a universal World Soul (an idea taken from Hinduism), a soul which subsisted in stars and in earthly objects alike. Our own, individual souls were but small particles of the great Universal Soul. However, as philosophers and *savants* proved unable to supply "truths" and detailed explanations about important matters (such as: What exactly happens after death?) men and women sought

---

[3] This arrogance about one's own language is one of the reasons for the political turmoil in the UK over Brexit. The British people have long been used to English being the universal language, or *lingua franca*; a situation that was certainly true for the century after the Battle of Waterloo, 1815. With the decline in the political status of the UK in the early 20th century and of the USA in the early 21st century, however, English is under threat as the *lingua franca*.

revelations beyond reason, through visions, and through mystical communication with the God-head itself. This individual search for experience of the Divine led to mysticism being practiced by individuals, and the search for salvation of an individual soul, personal salvation, which was radically different from the basis of Pagan belief systems.

Perhaps the syncretic religion that most blended physical and metaphysical concepts (that is blended matter and spirit, which the rationalist Greek philosophers had said could not be blended) was Pythagoreanism. The founder of this school was Pythagoras of Samos (c.570–c.495 BCE), whose political and religious teachings were well known in Magna Graecia and influenced the philosophies of Plato and Aristotle, and, through them, Western philosophy. Knowledge of the life of Pythagoras is clouded by legend. The teaching most securely identified with Pythagoras is metempsychosis, or the transmigration of souls, which holds that every soul is immortal and, upon death, enters into a new body. He may have also devised the doctrine of *musica universalis*, which holds that the planets move according to strict mathematical rules (Isaac Newton would have agreed with this idea) and thus resonate to produce an inaudible (to us) symphony of music.

In antiquity, Pythagoras was credited with many mathematical and scientific discoveries, in-cluding: Pythagoras' theorem; Pythagorean tuning; the five regular solids; the theory of proportions; and the sphericity of the Earth. It was said that he was the first man to call himself a philosopher, that is, a "lover of wisdom," and that he was the first to divide the globe into five climatic zones. Pythagoras influenced Plato, whose dialogues, especially his *Timaeus*, exhibit Pythagorean teach-ings. Pythagorean ideas on mathematical perfection also impacted ancient Greek art. His teachings underwent a major revival in the 1st century BCE among Platonists, leading to the rise of Neo-Py-thagoreanism and Neo-Platonism. Pythagoras continued to be regarded as a great philosopher throughout the Middle Ages and his philosophy had a major influence on scientists, or natural philosophers, such as Nicolaus Copernicus, Johannes Kepler, and Isaac Newton. Pythagorean sym-bolism led to early-modern European esotericism.

From its beginnings, Pythagoreans had regarded themselves as the keepers of a mystic tradition of knowledge and practiced initiatory rites—something that always attracts attention and new adherents. Their understanding of the laws of music and mathematics, as being the basis for the physical world, was presented as the fruit of revelation obtained from the most ancient of civilizations of which they were aware, the Egyptians. By the time of Pythagoreanism's second ap-pearance, however, Egyptian civilization had been eradicated by Greek civilization and then Latin conquerors. Ancient Egypt had become an enigma, a set of incomprehensible hieroglyphs. Yet there is nothing more fascinating than secret wisdom: one is sure that it exists and that it is hugely important, but one does not know what it is. In the imagination, therefore, it acquires exaggerated profundity. The language of Ancient Egypt, the hieroglyphs, naturally became the most ancient of languages—that of symbols.

For Saint Augustine of Hippo (354–430), as for nearly all the early fathers of the Church, Hebrew was the accepted primordial language. It was the language spoken before Babel, in the Garden between God and Adam. After the confusion induced by the fall of the Tower of Babel, Hebrew remained the tongue of the elected people. But Augustine was not interested in recovering its use. He was at home in Latin, by now the language of the empire, the church, and theology. Several centuries later, Isidore of Seville (560–636) found it easy to assume that, in any case, there were three sacred languages—Hebrew, Greek, and Latin. With this conclusion, the task of determining the language in which the God had said "*Fiat lux*," which had brought forth the visible universe out of nothing, became more difficult.

Figure 2.2: The Tower of Babel by Pieter Bruegel the Elder (1563). Image from: https://en.wikipedia.org/wiki/Tower_of_Babel#/media/File:Pieter_Bruegel_the_Elder_-_The_Tower_of_Babel_(Vienna)_-_Google_Art_Project.jpg.

There is one sense in which Saint Augustine did have a clear idea of a perfect language, common to all people. But this was not a language of words; it was, rather, a language made out of things themselves. He viewed the world as a book written by God's own hand. Those who knew how to read this book were able to understand the allegories hidden in the scriptures, where beneath references to simple earthly things (plants, animals, stories, etc.) lay hidden symbolic meanings. This Language of the World, instituted by its Creator, could not be read, however, without a key; it was the need to provide such a key that provoked a rapid outflowing of bestiaries, lapidaries, encyclopaedias, and *imagines mundi* throughout the Middle Ages. Many times in the last two millennia European culture has seized upon hieroglyphs and other esoteric ideograms, believing that fundamental truths are expressed in emblems or symbols,[4] and all we need do to return to the Golden Age is comprehend those hieroglyphs [1].

Between the fall of the Roman Empire and the early Middle Ages, new languages came into being, but without the nationalism of individual nations. It is believed that, toward the end of the 5th century, people no longer spoke Latin, but rather Gallo-Romanic, Italico-Romanic, early-Welsh (with Latin additions), or Hispano-Romanic, while *savants*, less gifted than previous generations of *savants*, continued to write Latin, bastardizing it ever further. They heard around them local dialects in which survivals of languages spoken before Roman civilization were grafted onto, or crossed with new vernaculars arriving with the barbarian invaders.

This age, characterized as Dark, seemed to witness a reoccurrence of the catastrophe of Babel: supposedly uncivilized and uneducated barbarians, peasants, artisans, the first Europeans—unlettered and unversed in official Latin-Greek culture—spoke a multitude of vulgar tongues of which official culture was unaware. It was the age that saw the birth of the languages which we speak today. European culture, and the cultures of those nations which started as European colonies, were all strongly influenced by these Dark age vulgar tongues. European critical culture begins with the reaction, often alarmed, to the explosive growth of the number of these tongues. Europe was forced at the moment of its birth to confront the drama of linguistic fragmentation, and European culture arose as a reflection on the perceived destiny of a multilingual civilization. Its prospects seemed uncertain; a remedy for linguistic confusion needed to be sought. Some *savants* looked backward, trying to rediscover the language spoken by Adam. Others looked ahead, seeking to create a rational language possessing the perfections of the lost language of power spoken in Eden. It was the latter that led to modern science, but the former path is still with us in the metaphysics of the search for the Theory of Everything (see Chapter 11).

2.2   A MYSTICAL LANGUAGE

The mystical approach to seeking the secrets of sacred texts is the Jewish esoteric tradition, known as the Kabbalah. In the 12th and 13th centuries, the Jewish communities of northern Spain and the south of France developed a tool for the textural analysis of sacred texts. The Kabbalah is a mystical technique for interpreting the first five Books of the Hebrew Bible, the *Torah*, and which regarded creation itself as a purely linguistic phenomenon. Beneath the letters in which the *Torah* is written today, the Kabbalist sought to identify the shape of what is termed the "eternal *Torah*," which had been created by God before He created the Universe, and which was believed by the Kabbalists to be the blueprint for Creation.

---

[4]   The best-selling *Foucault's Pendulum* (1988) by Umberto Eco is all about the search for supposed hidden knowledge, or hidden truth; and explains how this endless search, for something that likely doesn't exist has given rise to so many widely believed conspiracy theories.

The Kabbalist seeks to use the existing sacred text as an instrument. He knows that beneath the given text, beneath the familiar stories and events narrated in the *Torah*, there is another text which reveals a mystical and metaphysical reality. To uncover this mystical reality, and thus to come closer to the mind and intentions of the Divine, one must look beneath the literal narrative of the written text. Indeed, a Kabbalist would say that a sacred text can be read in four ways: (1) there is the simplistic or literal reading of the text; (2) there is an allegorical or philosophical manner in which to read the text; (3) the text may also be read hermeneutically (encompassing everything in the interpretative process including verbal and non-verbal forms of communication as well as prior aspects that affect communication, such as presuppositions, previous interpretations, the meaning and philosophy of language, and semiotics); and, finally, (4) the text may be read at the most profound level—at a mystical level.

Just as the Kabbalists spoke of the four levels of meaning in a sacred text, the poet Dante, a near contemporary of the Occitan Kabbalists (certainly of the greatest of the Iberian Kabbalists, Abraham ben Samuel Abulafia, the founder of the school of Prophetic Kabbalah, who was born in Zaragoza in 1240, and died sometime after 1291) and who knew of their ideas, considered that there are also four levels of meaning in poetry. In the *Divine Comedy* (*Inferno* IX 61–63), Dante speaks to the reader and tells him of the meaning that is hidden within the verses, "*O you whose intellects are sane and well,/ Look at the teaching which is here concealed/ Under the unfamiliar veil of verses.*" The reader is then led to understand the four meanings of the poem, that is, the literal meaning, allegorical meaning, moral meaning, and finally anagogical meaning. Dante's great poem is seen as a journey through and beyond life; an allegory about the stages of the soul's redemption; a warning and guide, and a prophecy of Divine things to come.

For the Kabbalist, language was a self-contained universe where the structure of the language represented the structure of physical reality. Thus, in contrast to the main schools of philosophy, in the Kabbalah, language does not represent the world merely by referring to it. If God created the world by uttering certain words or by combining certain letters, it follows that these elements were not representations of pre-existing things, but are the very forms by which the Universe was shaped and molded. The Divine Language was perfect not because it happened to reflect the structure of the Universe, but because it actually created the Universe. The Divine Language spoken by God, and used by Adam to name Creation, stands to the Universe in the same manner as the mold stands to the object cast from it.

But, if there are secrets about how the Universe came into being hidden in well-known sacred texts, why then is man not able to fully comprehend all the mysteries of the Universe and to work prodigies by uttering similar combinations of letters from the *Torah*? The Kabbalists say that the reason God hid the true meaning of the *Torah* after the fall of Adam; that is, He did not give man the correct order of the letters which compose the *Torah* because if He had given man the true *Torah* then anyone who could read this version would have the power to perform miracles.

To this end, the letters which form the *Torah* we have today have been considered, over millennia by *savants* and linguists, as the basis-functions of other combinations which have been used in an attempt to find the words of power which will empower the speaker to control Nature and to work miracles. This is the reasoning behind the meditation techniques (on the Divine Name) of Abraham Abulafia.

The *Torah* is interpreted as a mystical unity, whose primary purpose is not to convey a specific, simple, and literal meaning or story, but rather to express the immensity of God's power which is concentrated in His Name. The Kabbalists believe that the Name of God contains power, but at the same time it maintains and upholds the laws and harmonious order which pervade and govern the physical universe. Knowing the Name of God would enable man to penetrate the veil that separates the visible created world from the numinous; and just as God created the Universe by His speech, the man who knows the Name of Power could also directly influence Creation.

The Kabbalists, and those influenced by them also wished to read and fully comprehend the esoteric and apocalyptic books of the Bible, in particular, the Book of Daniel and the Revelations of Saint John. These Biblical texts were studied by Isaac Newton who was also searching for the Divine or mystical language. The Kabbalists and Newton (who possessed book on the Kabbalah) believed that Heaven and Earth were created by the uttering of the Name of God, and that the whole history and story of Creation were to be found in the gnomic utterances of the prophetic books of the Bible. The *Torah* was the source text in which to seek the power that had ordered Creation. Such concepts about names of power, and the power contained in such names, may seem more appropriate for the Classical or pre-Christian World, but names are very important, as we shall see in Chapters 13 and 14). Quintus Valerius Soranus (140–130 to 82 BCE) was a poet and Tribune of the Roman People at the end of the Roman Republic; he was crucified for revealing the secret arcane name of the Deity of Rome. To name something, or somebody is to know or understand that thing or that person, "*to name is to know.*" One did not reveal one's name lightly as we read in all our myths, fairy tales, and legends.

The mediaeval Kabbalists used combinatorial calculus (although they did not call it that) to combine various strings of letters in the established *Torah* as a means of seeking the Name of God. Intriguingly, at this same time and in exactly the same part of southern Europe, Ramón Llull was using very similar ideas to perfect his universal language of the philosophical and theological attributes of God. As we saw in Chapter 1, Llull was seeking a mechanical means of assisting the human memory and ingenuity in the association of combinations of characteristics and philosophical attributes, which is precisely what the contemporary Kabbalists were doing in their prayers and spiritual exercises.

The use of the Name of God to both affect and effect Creation is memorably expressed in the legends concerning the Golem, where a man could create a living but (importantly) an unthinking (that is, soulless) being from clay by use of the Ineffable Name of God. This story is beautifully described in Gustav Meyrink's novel, *The Golem* (published in serial form, 1913–1914), and in the magnificent silent movie, *Der Golem, wie er in die Welt kam*, based on this novel made by Paul Wegener in 1920.

Of course, what investigations such as the above demonstrate is that man has long been searching for something he believes he lost long ago, and the retrieval of which will bring about a new Golden Age for humanity. It does not matter if you are a theologian, an Eastern mystic, or a particle physicist; we are all looking for that lost something. The perfect language of man's innocence in the Garden. What was the language with which God conversed with Adam and in which God commanded Adam to name Creation? This perfect proto-language, (if it could be re-created) could be used to fully comprehend man's place in the Cosmos, and perhaps allow man to manipulate Nature itself. A great deal of the history of science, and nearly all pseudo-science is really the record of man's search for a simple language with which he could fully describe, comprehend, manipulate, and, perhaps, foretell or predict Nature, thereby allowing all men to comprehend all natural phenomena, whether known or as yet undiscovered.

2.3   FURTHER READING

1   A truly remarkable history of language; especially, the more esoteric aspects of that history: *The Search for the Perfect Language* (1995); Umberto Eco; Great Britain, Blackwell Publishers Ltd.

CHAPTER 3

The Mixing of Physics and Metaphysics to Create a Language of Curiosity

It is astonishing how many foolish things one can temporarily believe if one thinks too long alone.'

J.M. Keynes (1883–1946)

Languages are magical. They are the means of communicating to others our innermost secrets and thoughts, our desires and ideas. Such communication is not easy as we are not telepathic; we must construct sentences from the multitude of words in our memories. Such vocabulary is rendered into something that resembles our thoughts, via the rules of grammar, which are different for different languages but always serve the same purpose: to bring well-defined order out of chaos. But this process of verbal communication is rendered complex as no two people will describe something they both see in the same way; similarly, no two translators will render the same original text into exactly the same English text. Each of us brings to language, and to communication, our own experiences and limitations.

There is, perhaps, no better way of exploring man's abiding obsession with magic, with the occult, and with the hermetic arts than by looking at the ideas that have arisen about the creation of a single universal language. A language without ambiguity that would allow humanity to return to the Golden Age of simplicity, and the innocence of the Garden of Eden.

3.1   THE BIRTH PANGS OF MODERN SCIENCE

The 17th century was full of the reciprocal influences of mysticism on science, and science on mysticism; all mixed together by the solvent of philosophy, and inexplicable observations of Nature. It is said to be the time when astrology, alchemy, and magic yielded to Sir Isaac Newton, to scientific rationalism and universal laws. But is this really true? It is not more a case that the astrology and popular folk magic were still present in society, and widely accepted by all levels of that society when Newton published, the most influential of all textbooks of physics, his *Philosophiæ Naturalis Principia Mathematica* in 1687 and 1713? Indeed, the magic, religion, and superstitions of that period merely blended into the Newtonian view of the Universe; magic and proto-science were not immiscible fluids in the 17th century. Newton himself was not adverse to experimental investigations of alchemy (he wrote a million words on the subject), nor of attempting to comprehend Biblical prophecies, and he was a respected caster of horoscopes.

By the end of the 17th century, this mixture of popular and erudite beliefs had flowed together, and then been overlaid by something else. Today, it is as pointless to say that one set of beliefs gave way to, or was replaced by, another set of beliefs as it is to say that the Middle Ages ended in the mid-15th century, and were replaced by the Renaissance. No Age ever really ends, unless by the agency of a major catastrophe; the pre-existing Age is merely inundated by the succeeding Age. It is still there; just buried out of sight, and if some event causes the newer Age to be stripped away, the older Age will reappear as if nothing had happened in the intervening period. Today, belief in folk magic, superstition, and belief in the supernatural are still with us; many people today follow astrology, as did their ancestors in the time of Newton.

What is true to say is that the 17th century in Europe was a period of intense spiritual awakening. The continent was a bubbling cauldron of ideas and beliefs, where *savants* were rediscovering the ideas of the Neo-Platonists and Hermeticists. The Rosicrucian manifestos exploded into a world waiting and wishing for something to happen; into a world shaking off much of the sterile baggage of mediaeval Christian Scholastic dogma, and seeking new spiritual directions and dimensions. The Rosicrucian manifestos were a potent stimulus in a period when men were seeking renewal; a new way of looking at the world, which the geographical discoveries of the previous century had been shown to be larger and more complex than had previously been imagined. Some 17th-century *savants* poured over magical texts; others labored at forges, melting and distilling metals; other thoughtful, wise men sought to understand the stars, and to comprehend their silent, slow, sacred dances; and still others invented secret alphabets and universal languages attempting to better understand the ordering of Creation. All such men were looking not only to understand Nature, but also to control Nature.

In that period when magic and science mixed freely, everything was considered to be the hieroglyph of something else, and nothing was more lambent, more exciting, than a secret cypher. Galileo Galilee (1564–1642) was dropping weights from the Leaning Tower of Pisa and watching the isochronous oscillations of the chandeliers inside the nearby Cathedral of Pisa. In France, Cardinal Armand de Richelieu (1585–1642) was seeking to create a new political and economic order in Europe. All had their eyes peeled for signs and portents. All were searching for the unusual, for the new. The attractive pull of Newton's gravity and the oscillations of the pendulum became obsessions, and men not unnaturally reasoned that there must be something more; something quite different that lay behind, or perhaps above visible Nature. Another Italian *savant*, Evangelista Torricelli (1608–1647), inverted a long, glass tube filled with Mercury with the open end of the tube in a bowl of Mercury and invented the barometer with a *vacuum* at the top of the sealed column of Mercury, showing that man could recreate the primal nothingness or void. Torricelli may not have understood the physics of atmospheric pressure and how it changes from day to day, and from place to place; but he did know that he had captured a sample of the primordial nothingness from which the Cosmos had been created by God's command. To Torricelli's and Galileo's contemporaries, such experiments were as much about magic as they were about science, because there was no clear way of distinguishing between these two ways of looking at Nature (see Figure 3.1).

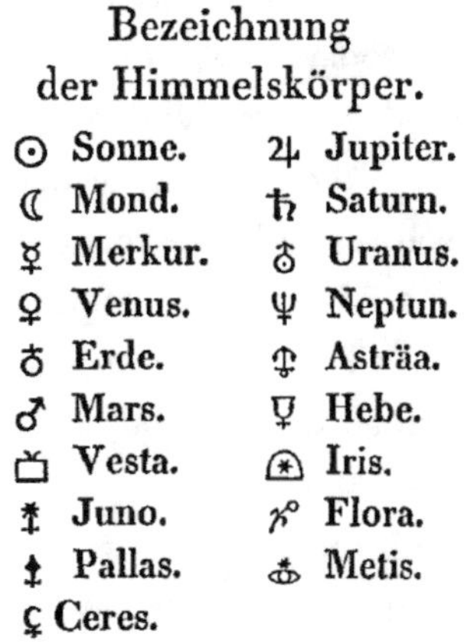
Bezeichnung
der Himmelskörper.
Sonne.
Mond.
Merkur.
Venus.
Erde.
Mars.
Vesta.
Juno.
Pallas.
Ceres.
Jupiter.
Saturn.
Uranus.
Neptun.
Asträa.
Hebe.
Iris.
Flora.
Metis.

Figure 3.1: A table of symbols for celestial bodies, from the *Berliner Astronomisches Jahrbuch* of 1850. This list is from the mid 19th century, yet the symbols of the planets used date from Antiquity. A planet symbol is a graphical symbol used in astrology and astronomy to represent a planet, including the Sun (*Sonne* in Figure 3.1) and the Moon (*Mond* in Figure 3.1). The symbols are also used in alchemy to represent the metals that are associated with the planets. The use of these symbols is based in ancient Greco-Roman astronomy, although their current shapes are a development of the 16th century. The International Astronomical Union discourages the use of these symbols in modern publications, and their style manual proposes one- and two-letter abbreviations for the names of the planets: Mercury (Me), Venus (V), Earth (E), Mars (Ma), Jupiter (J), Saturn (S), Uranus (U), and Neptune (N). The symbols of Venus and Mars are also used to represent the female and the male in biology and botany, following a convention introduced by Linnaeus (see Chapter X) in the 1750s. Even today, it is often difficult to separate the scientific and a more… magical description of Nature. (Image from: https://en.wikipedia.org/wiki/Planet_symbols#/media/File:Bezeichnung_der_Himmelskörper_Encke_1850.png.)

Indeed, at this time it is just about impossible to separate the world of magic and the world of science, where science is defined as the world of verifiable fact. *Savants*, who today are held up as paragons of rationalism, the standard-bearers of the scientific method, of rationalism, of mathematical and physical enlightenment—such as Isaac Newton (1643–1727), Robert Hooke (1635–1703), Blaise Pascal (1623–1662), René Descartes (1596–1650), and Francis Bacon (1561–1626)—turn up in, what we might call, the fog of superstition which clung to their age. Many of these men worked with one foot in the laboratory, and the other foot in the Kabbalah.[5]

The *savants* of the 17th century may have sought to understand the workings of the natural world, but they also dabbled in alchemy, biblical prophec, and astrological *cénacles*. Perhaps Newton arrived at his Universal Law of Gravitational attraction because he believed in the existence of occult forces, which recalled his life-long investigation into astrology (the influence of the orientation of the stars and the planets on a man's life) where he ended up proving that there is a force

[5] But as Joseph Needham (1900–1995), one of the wisest of 20th-century savants commented, "*Laboratorium est oratorium*"—the place where we do our experiments is also a place of prayer and contemplation.

between the planets and the stars, and that this force is in the nature of an attraction, and would be differently experienced by each and every infant.

Right up until his last days, Isaac Newton was working on a vast program that had obsessed him for half a century; to understand what he called the "*mystic language*" and to thereby fully understand the Divine Word as it came mysteriously from the mouths of the Prophets. A portion of his voluminous manuscripts on this work was published after his death in 1733, *Observations Upon the Prophecies of Daniel and the Apocalypse of Saint John*. Newton, who today is held up as the exemplar of rationality in a pre-scientific, barbarou,s and superstitious world, was not alone in this more mystical type of investigation. Many of Newton's contemporaries worked on projects of this type, which they hoped would lead to a renewal of Christianity and the spiritual enlightenment of man, not based on the views and authority of the Pope, or of a small group of isolated theologians and churchmen, but instead based on reason which would be accessible to all. They were seeking an unchanging universal standard of spiritual awareness; one not tied to artifacts kept in the vaults in Rome or Canterbury.

Today, many get embarrassed at the mention of magic, but even until the latter part of the 17th century practical magic was an accepted part of people's lives. For the scientists or natural philosopher of that period, the possibility of magic and magical acts and interventions was a fundamental presupposition. That the Bible contained fundamental truths about the place of man in the Cosmos was another presupposition, as was the possibility of identifying and using the Language of Creation.

The prevailing cosmology of that time was of an inanimate Earth, or elemental world subject to the influence of the heavenly bodies or stars and planets. This in itself was sufficient to encourage speculation about the astral origins for earthly phenomena, and to give rise to much speculation and lore about the astrologically derived properties of plants and minerals, and to speculate as to whether magic could be used to gain power over Nature, hence the similitude between the astrological/astronomical and alchemical symbols seen in Figure 3.1. The chemical/alchemical experiments of the magician, or natural philosopher, were often devised to identify and expose divine harmonies and correspondences between metals and planets. It also suggested that the magician might be able to find some means of tapping into the influence of the stars and diverting it to other purposes.

This is Neo-Platonism, which had for almost 2,000 years fostered a belief system which blurred the difference between matter and spirit, which had been the problem of philosophy in the West since it was first established in Ancient Greece; the Ancient Chinese were sensible enough never to have made such a separation. Instead of being regarded as an inanimate object, the Earth itself was deemed to be alive. (Today, we have returned to this essentially pantheistic view of Nature with the Gaia hypothesis of the self-regulating bio-sphere.) With its blurring of the boundaries between matter and spirit, Neo-Platonism also emphasized the influence of the human imagination upon the human body, of the mind upon matter and of words, incantations, and written charms upon physical objects. By the exercise of his imagination, and the use of magic, symbols, and incantations, the operator or magician (the magus) could transform either himself or his subject. The invisible powers of Nature were believed to be analogous to the invisible powers of the magnet, which could be clearly seen to act at a distance and penetrate matter with its invisible rays (see Figure 12.1). Since the world was a pulsating mass of vital influences and invisible spirits, it was only necessary that the magician should devise the appropriate technique to tap into them, and then to be able to work wonders. Today, we would say the world is a complex field of interacting virtual particles as in the quantum field theory view of the vacuum. Perhaps a single seamless manifestation arising from the never-resting quantum flux of the universal vacuum, that is, at the quantum level the phenomena of Nature are more closely coupled than the *savants* of the 17th century could ever have imagined.

In the 17th century, the Universe was believed to be peopled by a hierarchy of spirits thought to manifest all kinds of occult influences and sympathies, all linked together by invisible force fields through the Pythagorean Music or Harmony of the Spheres, and Newton's gravity. The Cosmos was thought of as an organic unity in which every part, or manifestation bore a sympathetic relationship to every other part. Even colors, letters, and numbers were endowed with magical or mystical properties. The investigation of such phenomena was the primary task of the early scientist or natural philosopher. Modern science merely grew out of our magical and religious view of the world around us; it was able to correctly answer a few more of the questions that man kept asking.

The project of finding the perfect, universal, divine or mystical language was something Newton followed all his life. It was a search for an unequivocal means of understanding the texts, which were the medium through which God communicated with man. First, Newton derived a mathematical language for formulating and predicting the dynamics of the heavens; then he would apply these techniques to the Bible and seek to tell us of the future. For Newton, numbers and equations were akin to those signs and clues by which the magicians or natural philosophers had first attempted to uncover the secrets of Nature. Newton regard Nature as a cryptogram created by God; as he put it, "*Numero ponderi et mensura Deus omnia condidit*" (*God created everything by number, weight and measure*). By pure reason, by hard work, by achieving wisdom through experimentation, this riddle could be solved by the dedicated savant. The question was one of discrimination; how do we tell the difference between the word of God and mere human words?

The 17th-century search for a universal language was not new, but was a natural wish in light of the gradual decline of Latin. Literature in vernacular languages became more prominent in the Renaissance, and during the 17th century learned works written by *savants* for other *savants* largely ceased to be written in Latin; and the rise of printing spread these vernacular texts. But what these *savants* observed was that the vernacular languages were not as logical or coherent as they would like for a medium of communication, which they wished to use to truly describe the world they saw around them. This desire for a reformed language, a philosophical language where words would perfectly describe objects, animals, and natural phenomena, and where the names of related objects, animals, and phenomena would be related to each other, led savants to consider again the Divine Language which Adam had used to name Nature (see Figure 2.1).

The German savant, and competitor with Isaac Newton, Gottfried Leibniz conceived of a "*characteristica universalis*" or universal character of a language; that is, an algebra capable of expressing all conceptual thought. This algebra would include rules for symbolic manipulation, what Leibniz called a *calculus ratiocinator* (a means of calculating the correct choice). His goal was to put debate, argument, and reasoning on a firmer basis by reducing much of it to a matter of calculation that men could grasp intuitively. This is, of course, what Ramón Llull had been seeking to achieve in the 13th century with his mechanical means of combining characteristics (see Chapter 1). There was no need to learn the endless lists of the current meanings of words, in the various vernacular languages; one could simply use algebra and mathematical manipulations on a small set of base concepts, or base units to describe everything in Nature. The *characteristica* would build an alphabet of human reasoning—a calculus of thought. It would be akin to throwing away the *Catalogue of Ships* found in Book 2 of Homer's *Illiad*, and instead looking for what it was you wished to know in the shield of swift-footed Achilles (see Chapter 1).

3.2   GOTTFRIED LEIBNIZ AND THE NATURE OF THE UNIVERSE

One of Leibniz's most important developments was that of binary arithmetic. Although Leibniz was not the first to conceive of this arithmetic, he did formulate it coherently and believed that the Ancient Chinese must have known about it, on the grounds that it was implicit in the *I Ching* (see Chapter 5).

The binary system is the simplest notation for numerals. Our decimal system has a choice of ten characters for each place (units, tens, hundreds, etc.). In the binary system, there are only two characters: one to designate an empty place, the other to indicate that the place is filled. Using the convention of 0 for empty, and 1 for filled, the system runs as follows: conventional decimal numbers (binary equivalents): 0 (0), 1 (1), 2 (10), 3 (11), 4 (100), 5 (101), 6 (110), 7 (111), 8 (1000), 9 (1001), etc. Although Leibniz was proud of his discovery of binary arithmetic, he did little with it. This is a pity, because if the arithmetic used by machines had been developed in the 17th century we might have been able to develop calculating machines long before the 19th century. However, the binary system came into its own only with the advent of semiconducting electronics in the 1960s (an electron is either in a site, or it is not in that site—see footnote on Page 7). As far as Leibniz was concerned, the greatest significance of his work on binomial numbers was metaphysical, as showing how the Universe could be seen as constructed out of a number.

Gottfried Leibniz's metaphysical account of the distinction between God and the visible universe had both mystical and moral repercussions. Leibniz held that the visible, created universe was distinct from God by virtue of its passive, material, and mechanistic nature. This led him to construe that matter is unreal, which means that the materiality of the world consists in an admixture of unreality, or not-being. God is a pure being: matter is a compound of being and nothingness. Leibniz elevated this idea into what he called a mystical theology, by developing the ideas of Pythagoras, who held that numbers (and the ratios of numbers; that is, proportions) were the ultimate realities, and that the Universe as a whole was harmonious. That is, it manifested simple mathematical ratios, like those of the basic intervals in music; the idea of the music of the spheres arose with Pythagoras back in the 6th century BCE. Leibniz's contribution was to make these numbers binary. He was thus able to say that just as the whole of arithmetic could be derived from 1 and 0, so the Universe was generated out of pure Being (God) and Nothingness. God's creative act was therefore at one and the same time a voluntary dilution of His own essence, and a mathematical computation of the most perfect number derivable from combinations of 1 and 0. Binary arithmetic was not merely a convenient notation for the hierarchy of all possible concepts, but it was the best way of representing their very essence, with 1 and 0 themselves functioning as the only basic concepts. Thus, Creation becomes a dialectic between something and nothing; between 1 and 0.

The 17th century produced many proposals for philosophical languages. The best known of these proposed philosophical languages, which we shall examine in Chapter 6, was that proposed by the Rev. John Wilkins in 1668 (*An Essay toward a Real Character and a Philosophical Language*), the classification scheme of which ultimately led to the Thesaurus.

The search for the universal or perfect language originally concerned an attempt to rediscover the primal matrix language. For many centuries, the leading candidate for this original or primordial language was Hebrew. Then in the 18th century, this search finally lost its utopian fervour, and its mystical component as the science of linguistics and the concepts of semiotics were born, and with them the Indo-European hypothesis of the origin of modern languages. For a long time, however, the idea of a primogenital language not only had a kind of historical validity—to rediscover the speech of all mankind before the confusion generated with the fall of the Tower of Babel—but it also entertained the minds of some of the greatest writers and thinkers of the Middle Ages and the Renaissance. This original language should incorporate a natural relationship between the words and the things we see around us in our world. Adam was told by God to name Creation, but how did he choose the names he used? The primordial language had a revelatory value, for in speaking it the speaker would automatically recognize the nature of the named reality, and it might even be possible for men to effect miraculous changes in Nature merely by speaking this original mother tongue.

A lot has changed over the last two millennia, but today science is the only possible universal language that man can ever know. Science is the only endeavor which creates an absolute authority to which we much all respond; science cannot be ignored, nor can it be defied, and the laws of physics are absolute. It is only through science that we can hope to understand how the Universe came into existence, and why it is the way we observe and measure it to be. And the basic or semantic primes of the modern language of science are the base units or quantities, which we use to describe all known physical and chemical phenomena; indeed, these are also the base units and quantities we would use to try and understand any new phenomena as yet unobserved.

This search for the perfect language, which is still ongoing in the physics community, could, perhaps, be considered ridiculous and pointless, but it could also be seen as arising from uneasiness, because people would like to find in the words we all use an expression of the way the world works and how history unfolds, and in this we have been regularly disappointed.

3.3   FURTHER READING

For all aspects of the rise of science from an occult and magical background, the works of Frances Amelia Yates (1899–1981) are all worth reading. She was an English historian who focused on the study of the Renaissance, and also wrote books on the subject of the history of esotericism. In 1964, Yates published *Giordano Bruno and the Hermetic Tradition*, an examination of Bruno, which came to be seen as her most significant publication. In this book, she emphasised the role of Hermeticism in Bruno's works, and the role that magic and mysticism played in Renaissance thinking. She wrote extensively on the occult or Neoplatonic philosophies of the Renaissance. Her books, *The Occult Philosophy in the Elizabethan Age* (1979), *The Art of Memory* (1966), and *The Rosicrucian Enlightenment* (1972), are major contributions, where the author deals with the supposed remoteness and inaccessibility of studies of magic and of the Hermetic arts. These volumes are available from Routledge Classics, Oxford, an imprint of Taylor and Francis.

A truly remarkable history of languages; especially, the more esoteric aspects of that history: *The Search for the Perfect Language* (1995); Umberto Eco; Great Britain, Blackwell Publishers Ltd.

CHAPTER 4

The Transformation of Magic and Mysticism into Science

So far as it goes, a small thing may give an analogy of great things, and show the tracks of knowledge.

Lucretius (99–55 BCE)

As mentioned in the Introduction, science, or rather the scientific worldview, was developed in early societies after the development of religion, and of a magical or mythological view of Nature. Men began to observe Nature with the desire to understand her better. They would have observed the progression of the Sun and the Moon: the ephemerides. This information would first have been memorized as literacy had yet to develop. As we have seen, eventually the long lists of observed phenomena and things would have been preserved, perhaps in painting, carvings, or the form of orally transmitted poetry or myth. And after many hundreds of generations, these myths, sculptures, and images would have formed the basis of a scientific interpretation of Nature; that is, first observation, then hypothesis, and then further confirmatory observation and, perhaps, observational proof of hypothesis. Let us now consider how this transformation to a scientific view-point came about. Given that Chinese civilization is our oldest and best documented record of how societies developed, evolved, and became interconnected; it is in Ancient China that we must look for the origins of proto-science (see Further Reading).

What was the main motive for the early natural philosophers, or Taoists of Ancient China, that compelled them to engage in the observation and study of Nature? The answer is straight-forward: to gain that peace of mind that comes from having formulated an hypothesis, however simple and provisional, about the most terrifying manifestations of Nature. Nature would have been seen as being all-powerful, and indifferent to the suffering of man. These ancient societies would have seen that when angered, Nature was able to easily perturb and destroy the fragile structure of human society. Nature still has this power, as the effects of global climate change upon our engineered structures gathers pace.

Whether the natural phenomena studied by the Taoists were earthquakes, volcanic eruptions, floods, storms, or the various forms of plagues and disease, at the beginning of the adventure of science man felt himself to be stronger, more secure, once he had differentiated and classified the phenomena that assailed him. This security was particularly the case when he could name those plagues and disasters. As we will see in many places in this work, "to name is to know"; if you can name something, you have power over it—or you think you do. The origin of the name given to the newly identified violent natural phenomenon, or plague would entail some character of the nature of that violent, destructive event. Thus, the men who observed Nature, and named and described the observed natural phenomena, would have formulated some naturalistic theory about the origin, and likely re-occurrence of those phenomena. This proto-scientific peace of mind was known to the Chinese as *ching hsin* (see Figure 4.1). The atomistic followers of Democritus and Epicurus in the West knew it as ataraxy (calmness, or peace of mind; emotional tranquillity).

Figure 4.1: Early Spring (a hanging scroll painting) by Guo Xi. Completed in 1072, this is one of the most famous pieces of Chinese art from the Song Dynasty (960–1279). The painting is a meditation on Nature. The poem in the upper-right corner was added in 1759 by the Qianlong Emperor; it reads: *The trees are just beginning to sprout leaves; the frozen brook begins to melt. / A building is placed on the highest ground, where the immortals reside. / There is nothing between the willow and peach trees to clutter up the scene. / Steam-like mist can be seen early in the morning on the springtime mountain.* Image from: https://en.wikipedia.org/wiki/Early_Spring_(painting)#/media/File:Guo_Xi_-_Early_Spring_(large).jpg.

The *Book of Master Chuang* by Chuang Tzu, dating from the Warring States Period (476–221 BCE) tells us that "*The true men of old had no anxiety when they awoke, forgot all fear of death and composedly went and came.*" These ancient men knew from their studies of Nature that there was an order behind the apparent violent indifference of Nature. For his part, Chuang Tzu, and other Taoists speak of "*Riding on the Normality of the Universe*," or on the "*Infinity of Nature*," and thus describe the sense of liberation which could be attained by those who could remove themselves from the petty squabbles of human society, and unify themselves with the great mystery of Nature; that is, to leave society and study Nature and become nature mystics (a beautiful term for a scientist). They had observed Nature, and had seen that it is possible for man to live in harmony with Nature, and not be the passive subject of Nature's more violent manifestations. The same confidence in the power of, and the security that comes from observation is found in the West in Lucretius (c.99–c.55 BCE). The *De Rerum Natura* speaks of observation and deduction (that is, modern empirical science) as the only remedy for the numerous fears of mankind. The following lines are repeated three times in Lucretius' poem:

> *These terrors, then, this darkness of the mind*
> *Not sunrise with its flaring spokes of light*
> *Nor littering arrows of morning can disperse*
> *But only Nature's aspect and her Law.*

(Translation by William Ellery Leonard)

In modern science, the relationship between the rational and the empirical is seen to be so obvious as to require no explanation. However, this was not always the case. To emerge into the light in Europe, the modern scientific method had to struggle against the dead-hand of formal, mediaeval scholastic rationalism. Even up until the early-17th century, the proper marriage of rational thought to empirical observations had not been consummated. At this time, it was considered in the ironic, but the Taoist words of Robert Boyle (1627–1691), the "father of chemistry" and author of the Sceptical Chymist of 1661, "*much more high and Philosophical to discover things a priore than a postiore.*"

The origins of modern science are to be found in the interaction of four tendencies: two on one side of the argument, and two on the other side. On one side, theological philosophy allied itself with Aristotelian scholastic rationalism to oppose those natural philosophers who wished to more fully understand Nature by observation. On the other side of the argument, were those natural philosophers who wished to use experimental empiricism to explore Nature. This later group would be experimentalists who reacted against Thomist scholasticism, and who found a powerful ally in mysticism. In the European Middle Ages, Christian theology, given its universal domination was on both sides of the arguments about the rise of the modern scientific method; all those for and against the development of the scientific method would have been believers. But while rational theology was anti-scientific; mystical theology, or a mystical or spiritual view of the Divine, in all its aspects proved to be pro-scientific. The explanation for this apparent contradiction is to be found in the nature of magic; that essential pre-scientific element from which science evolved. Rational theology was vehemently anti-magical (all those burnings of witches and heretics), but mystical theology tended to be more tolerant of magic and belief in magic; there was an affinity between mystical theology and Hermeticism in Europe, and that affinity arose from the study of Nature.

But the fundamental cleavage here was not between those who were prepared to use reason to understand Nature, and those who felt reason to be insufficient to understand Nature, but between those who were prepared to use their hands and those who refused to do so; between experimentalists and theoreticians. The Vatican theologians of the Inquisition who declined Galileo's offer to look through his telescope and to see for themselves that the Ptolemaic System was incorrect, were Scholastics and so they believed they were already in possession of sufficient knowledge about the visible universe. If after looking through his telescope, Galileo's findings agreed with Aristotle and with Saint Thomas Aquinas, there was no point in looking through the telescope; everything was already contained in the philosophy of the Angelic Doctor, Saint Thomas Aquinas. If the observations from the telescope did not agree with established dogma, they would have been dismissed, and condemned by the churchmen as magical and Neo-Platonic. As it turned out, there was a fundamental difference between Galileo's observations and the philosophy of Saint Thomas Aquinas; a difference that the Church of Rome did not accept until 1992.[6] Those Renaissance magicians, Nature mystics, and Neo-Platonists had discovered a new way of looking at Nature.

This explanation of the origins of the modern scientific method explains so much about the "less rational" interests of those earliest scientists. Why it was that men such as Isaac Newton, Robert Boyle, and Sir Thomas Browne (1605–1682) were interested in the Kabbalah and astrology, believing that such ancient mystical doctrines, the Hermetica, contained ideas of value to them in their new empirical studies. Typical of the attitude of these 17th-century experimentalists (the key here is that they were all experimentalists who 'got their hands dirty' in the laboratory) was that of the Flemish chemist, Jan Baptist van Helmont (1580–1644). Van Helmont was one of the founders of biochemistry; he was among the first to use a balance in quantitative experiments, he devised an early-form of thermometer, and he demonstrated the acid in the stomach and the neutralizing alkali of the duodenum. Yet for all this detailed interest in the reproducibility of experiments, van Helmont was also deeply anti-rational, displaying an almost religious empiricism. He attacked the hair-splitting formal logic, scholastic logic, which he felt had little relation to observable reality, but

---

6   In 1633, the Inquisition of the Roman Catholic Church forced Galileo Galilei to recant his theory that the Earth moves around the Sun. Under threat of torture, Galileo recanted. But as he left the courtroom, he is said to have muttered, "*all the same, it moves.*" 359 years later, the Church finally agreed. At a ceremony in Rome, before the Pontifical Academy of Sciences, Saint Pope John Paul II officially declared that Galileo was right. The formal rehabilitation was based on the findings of a committee of the Academy the Pope set up in 1979, soon after taking office. The committee decided the Inquisition had acted in good faith, but was wrong. The Inquisition's verdict was uncannily similar to cautious statements by modern officialdom on more recent scientific conclusions, such as predictions about greenhouse warming and climate change. The Inquisition ruled that Galileo could not prove "beyond doubt" that the Earth orbits the Sun, so they could not reinterpret scriptures implying otherwise.

merely trapped the mind in an endless circular argument. In truth, van Helmont was a European Taoist who believed in the need to observe Nature in order to understand Nature.

It may be said, therefore, that at the early stages of modern science in Europe, the mystical (nature mysticism, see Figure 4.1) was often more helpful than the rationalist approach, when it came to finding an explanation or a cause of an observed phenomenon. This situation exactly mirrors the rise of a scientific-like worldview among the philosophical scholars of Ancient China. Resting on the value placed on manual operations, that is, doing experiments rather than merely attempting to think of an explanation, and not testing that theory by experiment, men such as van Helmond and Isaac Newton were active laboratory workers as well as thinkers and writers. The equipment they built to undertake their experiments is still with us today. The Confucian social scholastics of Ancient China, like the rationalist Aristotelians and Thomists of mediaeval Europe nearly two millennia later, had neither sympathy for, nor interest in manual operations. Hence, practical science and magic were together driven into mystical heterodoxy. It is the association of nature mysticism and empiricism that is the foundation of post-Renaissance scientific thought in the West.

This same amalgam of empiricism and mystical theology, leading to a scientific worldview, can also be seen in Islam. The *Brethren of Sincerity* was an organization formed in Basra, Iraq, in about 950. Like the Chinese Taoists 1500 years earlier, and the Christian nature mystics of the 17th century, this semi-secret society had, at one and the same time, mystical, scientific, and political interests. The men who met in Basra acknowledged the existence of mysteries transcending reason, and believed in the efficacy of experimentation, particularly, manual laboratory experimentation in seeking to study those mysteries. All the *savants* involved in these early phases of the development of science recognized that effects may be brought about by specific manual operations without our being able to say exactly how or why; and they further believed that this unexplained data and information ought to be recorded and accumulated for future generations of savants. Their opponents, and often persecutors (as with their Christian and Chinese colleagues), believed that the nature of the Universe could be apprehended by ratiocination alone, and that quite enough information had already been made available to scholars, and that in any case, the use of the hands to do work of any kind was unworthy of individuals claiming to be scholars. The early proto-scientists were thus in a dilemma, for they could either set up a ratiocination of their own consisting of obviously inadequate theories and models, or rest on the thesis that "*there are more things in heaven and earth, Horatio, that are dreamed of in your philosophy*," and seek elucidation of the unexplained by further observation and study. Only cycles of experimentation and hypothesis would allow a resolution of this situation; thus, the modern scientific worldview was born.

There is of course a great difference between the nature mysticism or mystical naturalism, which triggered the creation of modern science (but which did not lead to the scientific triumphal-ism of the late 19th century) and other forms of mysticism, which are focused in purely religious contemplation, or meditation upon a God or gods. As is often said by theologians, religion is a belief in someone else's experience of the Divine, while spirituality is having your own experience of the Divine. The Ancient Chinese Taoists and the first European scientists learned that by looking at the world, and thinking about what it was they were observing, a larger number of individuals could gain a first-hand experience of the Divine; that is a *gnosis* (a revelation of knowledge) as the Ancient Greeks and early Christians would have described this glimpse of the truth, or this discovery.

All that nature mysticism asserts is that there is much in the Universe that transcends human reason, but since it required the empirical to the rational for comprehension it also implied that the sum total of incomprehensibility will diminish if men humbly (and without pre-conceived ideas) explore the occult properties and relations of things. Religious mysticism is very different; it dotes on an arbitrary uniqueness, and seeks to minimize or deny the value of investigations of the natural world. It is authority-denying mysticism, not rationalism, which at certain times in world history assists the growth of experimental science. And one may readily see why such upwelling of the scientific worldview often correspond with periods of social progress. As we read in the *Tao Te Ching*: "*Everyone recognises good as good, and thus what is not good… is also known.*"

4.1   FURTHER READING

In the sections of this present work where I discuss Ancient China, I will be making use of the magisterial *Science and Civilization in China* (published 1956, re-published 1975), Joseph Needham; Cambridge, Cambridge University Press. In this chapter, I make reference to Volume II of this multi-volume work, *The History of Scientific Thought*.

CHAPTER 5

The *I Ching* as a
Model of the Cosmos

The situations depicted in the Book of Changes are the primary data of life—what happens to everybody, every day, and what is simple and easy to understand.

Hellmut Wilhelm (1909–1990)[7]

The *I Ching*, also known as the *Book of Changes,* is an ancient Chinese divination text, and it is one of the oldest pieces of Chinese literature.[8] The text has been used for well over two millennia as a cultural reference, and has inspired ideas in religion, psychoanalysis, literature, and art. Indeed, the text has had a profound influence on western culture, but it originated as a divination manual in the Western Zhou Period (1000–750 BCE) of Ancient China. Then during the Warring States Period (475–221 BCE) and early Imperial Period, the *I Ching* was transformed into a cosmological text with a series of philosophical commentaries known as the *Ten Wings*. After becoming part of the *Five Classics* in the 2nd century BCE, the *I Ching* was the subject of scholarly commentary, and the basis for divination practice for centuries across the Far East, and eventually took on an influential role in western understanding of eastern thought. Various modern scholars suggest dates for the original text ranging between the 10th and 4th centuries BCE.

The form of divination involved in the *I Ching* is a type of cleromancy, which in the *I Ching* concerns the generation and interpretation of random numbers represented as figures termed hexagrams. The interpretation of the random readings, via the content of the *I Ching* is a matter of many centuries of debate, and many commentators have used the book symbolically, often to provide guidance for moral decision making informed by Taoist and Confucian ideals. The hexagrams themselves have acquired cosmological significance and become associated with other processes of change such as the coupled forces *Yin* and *Yang* and the five Chinese elements, *Wu Xing*. Many believe that the *I Ching* is a book containing an explanation of all the laws of physics, an expla-

---

7   Hellmut Wilhelm was the son of Richard Wilhelm (1873–1930), the German sinologist, theologian, and missionary, who is best remembered for his translations of philosophical works from Chinese into German, which in turn have been translated into other languages. His translation of the *I Ching* is still regarded as one of the finest, as is his translation of *The Secret of the Golden Flower*, both were provided with introductions by the Swiss psychoanalyst Carl Jung, who was a personal friend.

8   According to Google searches, the *I Ching* comes higher in the list of the most influential books than both the Old Testament and the New Testament.

nation of how everything is governed, and carries explicit directions on how men should conduct themselves in order to remain continually in harmony with these natural laws.

Within both modern physics and Eastern philosophy, it is believed that all natural phenomena in this world of change and transformation are dynamically interrelated. Emphasizing movement and change, Chinese philosophy had long ago developed concepts of dynamic patterns which are continually formed and dissolved again in the cosmic flow of the *Tao*. The *I Ching* has elaborated these patterns into a system of archetypal symbols, the so-called trigrams and hexagrams.

Ancient Chinese scholars contemplated the Cosmos in a way comparable to that of modern physicists, who with the advent of quantum mechanics introduced into their model of the Universe a psychophysical element. In the quantum view of Nature, the experimenter is an essential part of any experiment; as shown in the various interpretation of the (in)famous thought experiment by Schrödinger about a cat in a sealed box with a vial of poison. The observed microphysical event in an experiment necessarily includes the observer, just as much as the reality underlying the *I Ching* comprises subjective; that is, psychic conditions in the totality of momentary incidents and events. The 64 hexagrams of the *I Ching* become the instrument by which the meaning of the 64 different, yet typical, situations found in Nature can be determined. Therefore, for someone who regards the physical world in the same manner as the ancient Chinese scholars, the *I Ching* retains more than a slight attraction.

In its original structure, the *I Ching* is made up of eight trigrams, consisting of eight combinations of three lines of broken *Yin-Hsiang* (- -) lines and unbroken *Yang-Hsiang* (—) lines (see Figure 5.1). It is believed that these concepts have a cosmogonic significance. According to the Supreme Ultimate (Nothingness), a simple line symbolizing the positing of Oneness (—) produced the two modes *Yin* and *Yang* by splitting and filling of the lines. This creation ex-niho may seem strange, but as the Heart Sutra of Buddhism puts it:

> *O Sariputra, Form does not differ from Emptiness*
> *And Emptiness does not differ from Form.*
> *Form is Emptiness and Emptiness is Form.*
> *The same is true for Feelings,*
> *Perceptions, Volitions and Consciousness.*

And monotheism is not without its own creation ex-nihlo.

5.1   DETAILS OF THE *I CHING*

The Five Elements Theory (*Wu Xing*) has the same fundamental philosophy as the theory of the two coupled and inseparable forces *Yin-Yang*; that of continual evolution and balance. Each natural element (the five elements of Classical Chinese thought: wood, fire, earth, metal, and water compare with the four classical elements in the West: air, fire, water, and earth; see Table 7.1) has specific attributes that vibrate with their own frequency or energy. These elements interact with each other to affect the flow of energy in an individual's environment, in a positive or negative manner. *Feng shui* practitioners utilize the concepts of *Yin-Yang* and the Five Elements to balance competing energies in your environment. It is through the four *hsiang* that the eight trigrams are derived: each made up of combinations of three divided or undivided lines (see Figure 5.1). A summary of the properties and meanings of each of the eight trigrams is given in Table 5.1.

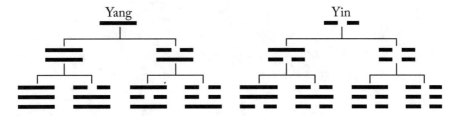

Figure 5.1: The eight trigrams derived from the Nothingness that also gave rise to the *Yin* and the *Yang*. The origin of the eight trigrams is the two coupled forces, *Yin* and *Yang*, and it is from these eight trigrams that the 64 hexagrams of the *I Ching* are derived.

Table 5.1: The names and the attributes of the eight trigrams. We saw earlier how the German *savant* Leibniz read the sequence of the trigrams as a perfect representation of the progression of binary numbers (000, 001, 010, 110, 101, 011, 111 ...)

Trigram	☷	☶	☵	☴	☳	☲	☱	☰
Name (Chinese)	Kun	Gen	Kan	Xun	Zhen	Li	Dui	Qian
Name	The receptive	Keeping still	The abyssal	The gentle	The arousing	The clinging	The joyous	The creative
Attribute	Devoted	Standstill	Danger	Penetration	Movement	Light giving	Pleasure	Strong
Image	Earth	Mountain	Water, clouds	Wind, wood	Thunder, wood	Lightening, fire	Lake	Heaven
Family relationship	Mother	Youngest son	Middle son	Eldest daughter	Eldest son	Middle daughter	Youngest daughter	Father
Binary numeral equivalence	0	1	10	11	100	101	110	111
Decimal equivalence	0	1	2	3	4	5	6	7

These eight trigrams are some of the most basic symbols of Eastern philosophy, representing transitional phases of Nature, and of human thought and psychology (see Table 5.1). They represent the maximum number of *Yin* and *Yang* relationships in sets of three. *Yin* and *Yang* forces are combined as *Yin/Yin*, *Yin/Yang*, *Yang/Yin*, and *Yang/Yang* combinations. These four combinations of forces are again divided to form the eight trigrams, and are said to be linked to the forces of Nature: Heaven and Earth, fire and water, thunder and wind, mountain and lake (as given in Table 5.1).

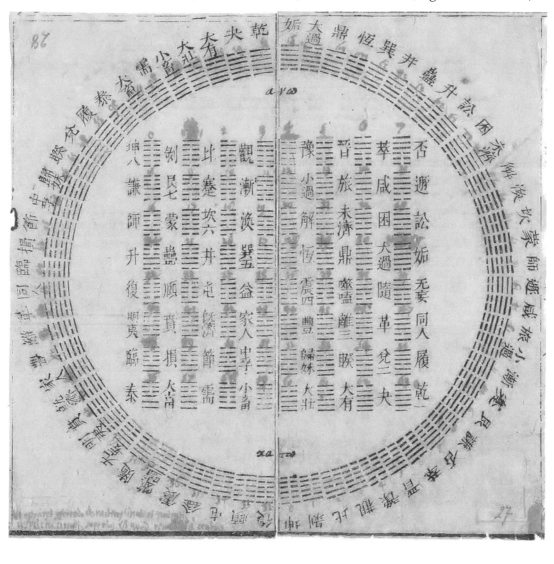

Figure 5.2: The 64 hexagrams of the *I Ching* (source: https://en.wikipedia.org/wiki/I_Ching#/media/File:Diagram_of_I_Ching_hexagrams_owned_by_Gottfried_Wilhelm_Leibniz,_1701.jpg).

Each trigram has its own name and property, and the trigrams are considered to represent all possible cosmic, natural, and human situations. They are associated with the phenomena of Nature, and the various possible situations in our social lives. They were also associated with the cardinal directions, and with the seasons of the year. Thus, the eight trigrams are often grouped around a circle in the natural order in which they were generated, starting from the top (where people in Asia have always located the south) and placing the first four trigrams on the left-hand side of the circle, the second four on the right-hand side. The objects or attributes thus symbolized by the eight trigrams are made to represent the constituents of the Universe, which form the basis of a cosmological system elaborated by the scholars of the Han Dynasty (206 BCE–220 AD) using the Five Element Theory.

By combining these trigrams as a symmetric matrix, a total of 64 combinations is obtained; known as the 64 hexagrams (shown in an ancient Chinese manuscript in Figure 5.2). The 64hexagrams are traditionally arranged in 2 patterns: (i) a square of 8 x 8 hexagrams and (ii) a circular sequence showing the same symmetry as the circular arrangement of the trigrams; both are seen in Figure 5.2. The 64 hexagrams are the cosmic archetypes on which the use of the *I Ching* as an oracle book is based. For the interpretation of any hexagram that may arise in the initial selection or divination, the various meanings of its two trigrams have to be taken into account.[9]

In the *I Ching*, the trigrams and hexagrams represent all possible combinations generated by the dynamic interaction of the forces *Yin* and *Yang*, and are reflected in all cosmic and human interactions. All things and situations are in a state of continual transition: one changing into another, the solid lines pushing outward and breaking in two, the broken lines pushing inward and growing together. Therefore, the 8 trigrams, together with the 64 hexagrams are deemed to represent all the possible situations and temporal mutations of phenomena in the Universe. The *I Ching* is believed to describe a system of metaphysics relating the Universe and natural phenomena, as functions of the time of day, seasons, weather, family relations, personal relations, etc.

5.2   DIVINATION WITH THE *I CHING*

To use the *I Ching*, an individual must first embark upon a process of divination; that is, they must allow themselves (in what they do) to be open to the influence of Nature. The person who has a question to ask of Nature must first invoke the forces of Nature; for example, by tossing a set of coins, or drawing lots to generate a set of results that may then be interpreted via the traditional text of the *I Ching*, which is the result of more than two millennia of interpretation of such divination experiments. The process of consulting the *I Ching* as an oracle involves determining the hexagram by a method of random number generation, and then reading the text associated with that hexa-

---

[9]   The Internet is full of sites detailing how these interpretations are made.

gram. Confucius said that one should not consult the Oracle for divination until one has passed the age of 40. Those studying the *I Ching* should also be free of compulsion; that is, repeatedly asking the same question in hope of either a different/better answer, or further enlightenment as to the meaning of the answers one first obtains.

Hexagrams were traditionally generated by the casting of yarrow stalks (*Achillea millefolium*). The stalks must be cut and prepared, being plain, lacquered, or varnished. Fifty yarrow stalks are used, though one stalk is set aside at the beginning and takes no further part in the process of consultation, or divination; this is the Wu Chi—the unchanging ground of being. The remaining 49 stalks are roughly sorted into 2 piles, and then from each pile 1 stalk is initially removed, then the pile is "cast off" in lots of 4; that is, groups of 4 stalks are removed. The remainders from each half are combined (traditionally placed between the fingers of one hand during the counting process) and set aside, with this process being repeated twice; that is, a total of three times. The total stalks in the remainder pile will necessarily (if the procedure has been followed correctly) be 9 or 5 in the first count and 8 or 4 in the second. Nine or 8 is assigned a value of 2; 5 or 4 assigned a value of 3. The total of the three passes will be one of only four possible values: 6 (2+2+2), 7 (2+2+3), 8 (2+3+3), or 9 (3+3+3); that count provides the number of the first line of the hexagram. When three successive changes produce the sum 3+3+3=9, this makes the old *Yang*, i.e., a firm line that moves. The sum 2+2+2=6 makes old *Yin*, a yielding line that moves. Seven is the young *Yang*, and eight the young *Yin*; they are not taken into account as individual lines. The 49 stalks are then gathered and the entire procedure repeated to generate each of the remaining 5 lines of the hexagram. (Each succeeding line is written above its predecessor; that is, the first line is at the bottom of the stack of lines, and the final, sixth line is at the top.)

During the Eastern Han Dynasty (1st century AD), there were two schools of interpretation of the *I Ching*. The first school, known as New Text Criticism, was more egalitarian and eclectic, and sought to find symbolic and numerological parallels between the natural world and the hexagrams. With the fall of the Han Dynasty, *I Ching* scholarship was no longer organized into systematic schools. One of the most influential writer of this period was Wang Bi (226–249), who discarded the numerology of Han commentators and integrated the philosophy of the *Ten Wings* directly into the central text of the *I Ching*, creating such a persuasive narrative that earlier Han commentaries were no longer deemed important. By the 11th century, the *I Ching* was being read as a work of intricate philosophy, as a starting point for examining metaphysical questions and ethical issues. Cheng Yi (1033–1107), founder of the Neo-Confucian Cheng–Zhu school, read the *I Ching* as a guide to moral perfection. He described the text as a way for ministers to formulate honest political opinions, and so avoid factionalism, to root out corruption, and to solve problems in government. The contemporary scholar Shao Yong (1011–1077) rearranged the hexagrams in a format that resembles modern binary numbers, although he did not intend his arrangement to be used mathematically. This arrangement, sometimes called the binary sequence, is the format that later inspired Gottfried Leibnitz; when the text had been translated and published in Europe by the Jesuits.

Gottfried Leibniz, who was corresponding with the Jesuit missionaries in China, wrote the first European commentary on the *I Ching* in 1703, arguing that it proved the universality of binary numbers and Theism, since the broken lines, the "0" or nothingness, cannot become solid lines, the "1" or oneness, without the intervention of God (see Page 35). This mystical interpretation was criticized by Georg Wilhelm Friedrich Hegel, who proclaimed that the binary system and Chinese characters were "*empty forms*" that could not articulate spoken words with the clarity of Western alphabets. In their discussion, *I Ching* hexagrams and Chinese characters were conflated into a single foreign idea, sparking a dialogue on Eastern and Western philosophical approaches to questions such as universality, semiotics, and the nature of communication.

Following the *Xinhai* Revolution of 1911, the *I Ching* was no longer part of mainstream Chinese political philosophy, but it maintained a huge cultural influence as China's most ancient text. Borrowing back from Leibniz, modern Chinese writers offered parallels between the *I Ching* and subjects such as linear algebra and logic in computer science, seeking to demonstrate that ancient Chinese thought had anticipated Western discoveries. The Sinologist Joseph Needham (1900–1995) took the opposite viewpoint, arguing that the *I Ching* had actually impeded scientific development in China by incorporating physical knowledge into its metaphysics. The psychologist Carl Jung took a great interest in the possible universal nature of the imagery of the *I Ching*, and he introduced an influential German translation by Richard Wilhelm by discussing his theories of archetypes and synchronicity. The book had a notable impact on the 1960s counterculture, and on 20th Century writers and musicians such as Philip K. Dick, John Cage, Jorge Luis Borges, and Hermann Hesse.

5.3   FINAL COMMENTS

In the initial phases of the transformation of magic into science in Europe, the mystical approach to Nature was often more helpful to *savants* and those seeking to comprehend what they saw around them than the theoretical or rationalist approach. After all, if one begins with a rationalist approach to the investigation of Nature, one will soon end up doing experiments to test and thus extent theoretical models. And experiments are often difficult; they may not work for a whole host of reasons, and they may well be difficult and expensive. On the other hand, mystical interpretations of what one encounters in the natural world are cheap, and require no testing, only a vivid imagination and a knowledge of ancient history to provide the ancestral deities responsible for whichever natural phenomenon is being considered. However, experience tells us which of these two approaches yields the most useful, that is, reproducible results. In the British Isles, the Anglo-Irish alchemist and proto-scientist Robert Boyle paid considerable attention to this problem of resolving the use of theory versus experiment (even mystical theories versus rationalist theories) in his publication of 1661, *The Sceptical Chymist*. Boyle came down on the side of practical experimentation as being the ultimate, the acid test, for all speculation about Nature.[10]

Since each of the 64 symbols, or hexagrams, of the *I Ching* came, in the course of the centuries, to have an abstract signification, such a reference was naturally alluring and saved all necessity for further thought and any experimental investigation. The technique of the *I Ching* resembled in many aspects the astrological pseudo-explanations of Nature and man's destiny of pre-Renaissance Europe, but with the greater complexity (64 hexagrams as opposed to 12 Houses of the Zodiac); abstractness of symbolism gave it a deceptive sophistication.

The 64 symbols, or hexagrams, in the system provided a set of abstract concepts capable of subsuming a large number of the events and processes, which any investigation is bound to find in the phenomena of the natural world. It has been said that the *I Ching* supposes a kind of translation of all natural phenomena into a mathematical language by means of a set of graphic symbols, the germs of what the German philosopher and mathematician Gottfried Leibniz would have called a *universal language* or a *universal character*, thus constituting a dictionary capable of permitting men to read Nature like a book whether with intellectual, or practices aims in view. This is, of course, as much about true science as it is about astrology. Furthermore, the *I Ching* brings us back to the illusory realms of numerology, where numbers are not the empirical and quantitative servants of science, but a straightjacket into which theories have to forced to fit our pre-conceived ideas. To paraphrase Jung, the *I Ching* has more to do with synchronicity than with physics. Yet for all its flummery and quackery, and lack of anything other than a statistical success rate, it is a technique which is still hugely followed (much like astrology).

What seems to show through when one looks at the ideas of Taoism, and other similar thoughts about the origin and usefulness of the *I Ching* in early eastern natural science, is the effort made by the School of Naturalists and the Han Dynasty Confucians to use the figures made by the long and the short strokes; that is, the 64 hexagrams, as a comprehensive system of symbolism containing, in some way all the basic principles of all natural phenomena. That is, to construct a proto-language of science; even if it was a symbolic representation of this language. Like the Taoists, the naturalists who invoked the *I Ching* to comprehend the world were looking for peace of mind, as opposed to the worry of trying to learn long lists of things and phenomena, and forgetting some part of that list.

It is likely that a similar argument can be made to account for the central importance of astrology in the Mesopotamian civilization. The Houses of the Zodiac and the Sun, Moon, and planets formed a sufficiently complex system that permitted a range of correspondences to be constructed and maintained. If then one projects these theoretical, mystical interconnections onto ob-

---

[10] Happily, experiment won out in the end, although theoretical physicists still have an exalted status in the physics community.

servations of Nature, one does have a system of sufficient flexibility to explain some part of Nature. But of course, this is a mystical theoretical model of Nature; one could say a mythological model. And this mystical model yielded to rational experimentation and evidence in the modern world, in both the East and the West.

5.4   FURTHER READING

The Internet is crowded with sites providing information about the *I Ching*, and about the interpretation of results derived from divinations using the *I Ching*. As for recommendations for further reading, I suggest:

1   The American physicist and ecologist Fritjof Capra (born 1939) has explored the parallels between modern physics and Eastern Mysticism in *The Tao of Physics* (1975); Boston, Shambhala Publications, Inc.

2   *I Ching* translated by Richard Wilhelm (2003); London, Penguin Books. This book has a forward by Carl Jung.

CHAPTER 6

Natural Philosophy

How many angels can dance on the head of a pin?

(A standard question for students of Scholasticism in the 13th century)

In our consideration of the concept of a perfect language, with which and through which man might truly appreciate and, perhaps, control Nature, we must now leave behind the fascinating but strangely exotic mixture of magic and science that had characterized the pre-scientific world, and examine the advent of the *a priori* philosophical language. The members of this new group of 17th-century seekers after a simpler, more perfect language were not magicians or Hermeticists, but *savants* and natural philosophers who sought a simple, but logical language which could eliminate the concepts and formalisms, which had previously clouded the judgment of men, and which had kept all men from fully and rapidly embracing the progress of science and technology.

Jan Amos Komensky (1592–1670; he used the Latinized form of his name as Comenius) was a Protestant mystic from Bohemia. Although inspired by religious ideals, he is considered to be one of the first *savants* who as part of their investigation of Nature tried to formulate a more perfect language to describe his observations, and to allow him to transmit his observations to other *savants*. In his *Pansophiae Christianae III* (1639–1640), Comenius advocated a reform of the commonly used vernacular languages to eliminate the rhetorical and figurative use of words, which he regarded as a source of ambiguity and confusion. The meaning of the words that remained should then be fixed, with one name for each thing; this, he believed, would restore words to their original meaning.

Although Comenius was never to construct his reformed, plain language, he had broached the idea of a universal language that attempted to overcome the political and structural limitations of Latin, which was still being used as a sort of universal language in Catholic countries. (Comenius came from the non-Catholic part of Central Europe, which from 1618–1648 was fighting for its existence in the Thirty Years War.) Comenius proclaimed that the lexicon of the new philosophical language would reflect the composition of reality, and every word in it should have a fixed, definite, and univocal meaning. Every idea should be represented by one and only one expression, and these definitions and expressions should not arise from an individual author's fancy or imagination, but should represent only things that existed. Comenius wished to create a utopian language that would describe the fixed, unmoving connections of every element of Creation; but he recognized that it would not be a vehicle for the creation of great literature.

The utopian ideas of Comenius would have necessitated a prodigious ability at memorizing all the new words and the new meanings. But this was exactly the type of problem that had inspired Ramón Llull to invent his *Ars combinatoria*. The French philosopher René Descartes saw where the real problem lay with such new philosophical languages. In order to avoid having to memorize and learn how to use the new fundamental or primitive names Descartes conjectured it only would be necessary for these to correspond to an order of ideas or thoughts which had a logic of their own akin to that of numbers. That is, that it was through the medium of mathematics and mathematical logic that the new universal language would eventually come into being. Descartes pointed out that if we can count, we are able to generate an infinite series of numbers without needing to commit to memory the whole set of all possible numbers. But this problem coincided with that of discovering a philosophy capable of defining a system of clear and distinct ideas. If it were possible to enumerate the entire set of simple ideas from which we mentally generate all the complex ideas of which it is possible to conceive, and if it were further possible to assign to each idea a character, as we do with numbers, we might be in a position to manipulate them with a mathematics of thinking, or a calculus of thought, while the words of natural languages evoke only confusion. This was the idea that was pursued in Germany by Gottfried Leibniz. What we have here in the first half of the 17th century is a statement about the essential properties of a computer language, three centuries before the invention of the computer.

In 1654, the English clergyman, alchemist, and astrologer John Webster (1610–1682) wrote his *Academiarum examen*, an investigation and attack on the academic world, which Webster felt had not given sufficient attention to the problems of creating a universal language. Like many English contemporaries of Comenius, Webster was influenced by the Bohemian's ideas. Webster foresaw the birth of a *"Hieroglyphical, Emblematical, Symbolical and Cryptographical learning."* Describing the general utility of algebraic and mathematical signs, numbers, and equations, Webster went on to say that *"the numerical notes which we call ciphers, the Planetary Characters* [the internationally known symbols for the planets, see Figure 3.1], *the marks* [the well-known alchemical emblems, see Figure 1.2, and Table 7.1] *for minerals and many other things in Chymistry, though they be always the same and vary not, yet are understood by all nations, and when they are read, everyone pronounces them in their own Country's language and dialect."*

John Webster was attempting the synthesis of mathematics and alchemical and astrological symbolism (today we would rather say chemical and astronomical nomenclature). He went on to say that such a symbolic language would be the true philosophical or universal language. Webster was something of a controversial figure in his own life; an Anglican clergyman who supported the Parliamentary cause in the Civil War, who was openly an alchemist and astrologer and who was sceptical about witchcraft. Yet, in Puritan England, this chaplain to the Parliamentarian army produced a work that was at the center of the 17th century's magico-scientific Hermetic tradition (see Figure 6.1), which also produced the astrology, mathematics, and Adamic language of Dr. John

Dee, the eminent mathematician and astrologer to Queen Elizabeth I, and the Angel languages and alchemy of Robert Fludd.

Figure 6.1: Hermes Trismegistus, a floor mosaic in the Cathedral of Siena (image from: https://en.wiki-pedia.org/wiki/Hermes_Trismegistus#/media/File:Hermes_mercurius_trismegistus_siena_cathedral.jpg). The mythic personality, Hermes Trismegistus, is associated with the Greek god Hermes and the Egyptian god Thoth. Greeks in the Ptolemaic Kingdom of Egypt recognized the equivalence of Hermes and Thoth, and the two gods were worshiped as one in what had been the Temple of Thoth in Khemenu, which was known in the Hellenistic period as Hermopolis. But the "personality" of this cultural mix of Ancient Egyptian and Greek gods became overlaid with something more. Hermes, the Greek god of interpretive communication, was combined with Thoth, the Egyptian god of wisdom. This multi-faceted deity thus became a god of wisdom. And it was as a source of all wisdom that he became known to the Neo-Platonists in the early centuries of the Christian era, particularly, in the Egyptian metropolis of Alexandria. As a divine source of wisdom, Hermes Trismegistus was credited with many writings, which were reputed to be of immense antiquity. Early Christians and Neo-Platonists were under the impression that the Egyptians had 42 sacred writings by Hermes, writings that detailed the training of Egyptian priests. These Hermetica are a collection of papyri containing spells and induction procedures for new adepts. The dialogue called the *Asclepius* (after the Greek-god of healing) describes the art of imprisoning the souls of demons, or of angels in statues with the help of herbs, gems, and odors, so that the statue could speak and engage in prophecy. This corpus of ancient wisdom was, however, merely a compilation of facts and a list of old observations. There was no underlying coherence, and all context had been lost. We are back with Homer's *Catalogue of the Ships*, but the literary and historical context had been entirely lost. Yet not only did this list lead to science, but it also influenced Christian dogma.

Not surprisingly, Webster was attacked and his ideas were ridiculed by contemporaries, however, his ideas were within the development of a universal, symbolic language based on mathematics and symbols, and not based upon phrases and the rules of grammar needed to try and keep order among the rapidly accumulating words of even a reformed language. The more mystical Hermetic ideas of Webster were denounced by John Wilkins (1614–1672), another Anglican clergyman and natural philosopher who was quite prepared to accept that a new language could be elaborated in which letters of the alphabet stood for mathematical quantities. But the critics of Webster argued that the only real character of which Webster spoke was actually the natural language of which the Kabbalists and Rosicrucians had sought for vainly in Hebrew. In spite of these mutual criticisms, the projects of the religious mystics did have something in common with those of the natural philosophers. The 17th century was full of reciprocal influences of mysticism on science and science on mysticism, all mixed together by the solvent of philosophy and observations of Nature that were at that time inexplicable. However, as we move toward the ideas of John Wilkins, we finally move away from the search for the lost language of Adam, and move to the secular world, which would centuries later lead to linguistics, semiotics, computer codes and modern science.

The first serious attempt at producing a systematized universal language based on philosophic principles was due to John Wilkins. He was a polymath who became the Bishop of Chester and the brother-in-law of Oliver Cromwell; he was one of the pre-eminent scientific innovators of that period. Wilkins assisted in the founding of the Royal Society of London. He was one of the creators of a new natural theology which attempted to be compatible with the science of the time, attempting to synthesize natural philosophy with the theology and dogma of the Church of England.

In 1668, Wilkins published his *Essay toward a Real Character and a Philosophical Language*, where he attempted to create a universal language to replace Latin as an unambiguous means of communication with which international scholars and philosophers could communicate. The *Essay* also proposed ideas on weights and measure which were similar to those which would later be found in the Metric System of 1795. In particular, Wilkins suggested that a more perfect system of weights and measures, a universal system of weights and measures could be generated by using the decimal metric system based upon a single universal measurement. In essence, Wilkins proposed that an entire system of units could be based on a single natural dimension, or *universal measure* (see Chapter 9).

John Wilkins spoke of a single *universal measure* upon which all other measures could be based, and from which all other measures could be derived by mathematics. Wilkins' *Essay* was translated into Italian in 1675 by Tito Livio Burattini (1617–1681), who translated Wilkins' phrase *universal measure as metro cattolico*, thereby introducing the familiar modern word, meter, or "measure." Tito Livio Burattini was a true Renaissance man, as interested in architecture and the designing of scenery for theatrical spectacles as in measurement science and mathematics. It was toward the end of his life in 1675 that he published his *Misura universale* (*Universal Measurement*) in Italian where he described the ideas of John Wilkins (whose essay had been published by the Royal Society of London in 1668). Burattini was one of the first European *savants* to make a detailed survey of the architecture of the Great Pyramid of Giza. Indeed, on his expedition to Egypt he was accompanied by the English mathematician John Greaves, who went on to become a professor of geometry at Gresham College, the forerunner of the Royal Society of London. Interestingly, the measurements of the Great Pyramid made by Greaves were later used by Isaac Newton in his studies of Biblical Prophecy and in a calculation Newton made of the circumference of the Earth.[11] But as we have seen, the 17th century was characterized by this curious mixture of science and magic, physics, and metaphysics.

John Wilkins wished to create a universal language, primarily to facilitate international communication among scholars, but he also envisioned its use by diplomats, travelers, and merchants. Wilkins' idea was to create a family of symbols corresponding to a complex classification scheme, which was intended to provide elementary building blocks which could be used to construct every possible object and idea. The *Real Character* is not a written representation of speech. Instead, each symbol directly represents a concept, without there being any way of speaking, or vocalizing it; each reader might, if he wished, give voice to the text in his or her own tongue. Later in his *Essay*, Wilkins introduces his *Philosophical Language*, which assigns phonetic values to the *Real Characters*, should it be desired to read the text aloud without using any of the existing natural languages.

In this universal language, each word defines itself. Descartes had already noted in 1629 that using the decimal numbering system it was straightforward to name all the numbers up to infinity, and to write them in a new language—if one were so disposed. Descartes went on to suggest the creation of a language similar to this numbering system, but a general language, organizing and covering all human ideas. In 1664, Wilkins started to work on this task.

John Wilkins divided everything in the Universe into 40 *categories* or *genera*, these being further subdivided into two hundred and 51 *characteristic differences*, which were subdivided into 2,030 *species* which appear in pairs. After depicting Nature in tables that occupy 270 folio pages, Wilkins began the construction of his philosophical grammar. Wilkins assigned to each *genus* a symbol consisting of a monosyllable of two letters; the *characteristic differences* are expressed by the consonants B, D, G, P, T, C, Z, S, N and the *species* by the addition of another letter; seven (Latin and Greek) vowels and two diphthongs. For example, *De*, signifies an *element*; *Deb*, the first *difference*, which according to Wilkins' Tables is fire; and *Deba* will denote the first *species*, which is flame. *Det* will be

---

[11] It was believed in the 17th and 18th Centuries that the Great Pyramid of Giza was a structure associated with magical and occult rites (Hermes Trismegistus again). In the same way that 18th Century surveyors were measuring the Meridian through Paris to define the universal measure of length (our modern meter), it was thought that the Ancient Egyptians had defined the near universal unit of length measurement in the Ancient world, the cubit from the base of the Great Pyramid (each base side is 440 cubits or 230.4 m) which was held to be 1/500 of one degree of the Earth's circumference (222.639 m).

the fifth *difference* under that *genus*, which is appearing meteor, and *Detα* the first *species*, which is rainbow. The words in the analytical language of John Wilkins are full of meaning and information; however, there is a great deal of arbitrariness. *Debα* signifies flame, because *α* designates a species of the element fire. If we replace *α* with *a*, we obtain a new symbol, *Deba*, which according to Wilkins' Tables designates comet; *Deba* and *Debα* are related but different.[12]

The "words" of Wilkins' analytical language contain no arbitrary symbols. Each letter in the analytical language has significance and meaning, in the same manner that the text of Holy Scripture has various levels of meaning for the Kabbalists, and the long numerical sequences which regulate our lives such as social security numbers; telephone numbers; computer access codes; bank account numbers contain all that there is to know about each and every one of us. (For example, the IBAN bank code has 22 characters and the SWIFT code has 11 characters—these numbers contain the potential for vast numbers of combinations allowing all humanity to be numbered and sorted, and then monitored.)

One could learn Wilkins' language without knowing that it was artificial. Then later, one could be led to discover that it was also a universal key and a secret encyclopaedia. Not surprisingly, Wilkins' language was not ideal, however, it was an extraordinary achievement for its time. But then, the impossibility of truly representing the entire scheme of the Universe should not and cannot stop us from planning human models, even though we are conscious that they are, at best, preliminary, and will, of necessity contain arbitrary and conjectural elements. The reason for this lack of perfection in any human attempt at categorising all knowledge is because we do not truly know what the Universe is, or indeed, where it came from, and why it is here. And it was man's speculation as to the origin and purpose of the visible world, that is, in God's secret dictionary that started him off on his search for the perfect or universal language. We have perhaps been going around in circles, but coming to new wisdom at the completion of each circuit. It is with the advent of a more perfect analytical language, that is, the modern physical sciences that man has begun to truly penetrate the Divine scheme of the Universe.

The analytic language of Wilkins is an example of a scheme intended to order all knowledge and relieve our memories of much unnecessary work. The word salmon, for example, tells us nothing, but *zana*, the corresponding word in Wilkins' classification, defines (for one versed in the 40 categories of *genera*, and the *differences* and the *species* of those *genera*) a scaled river fish with reddish meat. It was a century later that the Swedish botanist Carl Linnaeus (1707–1778) would adopt the familiar binomial system of nomenclature for all creatures (see Chapter 13); the extant and the fossil. Man being *Homo sapiens* ("thinking man") in the Linnaean classification, and the Atlantic salmon is *Salmo salar* ("leaping salmon").

---

[12] Umberto Eco gives a clear and fascinating description of the structure and use of Wilkins' Analytical Language in his *The Search for the Perfect Language*.

In the same way that Ramón Llull sought to use mechanical or logical devices to construct all possible combinations of the attributes of God (philosophical, theological and personal), thereby avoiding the tedious necessity of preparing *ab initio* lists, which would be very long and probably incomplete, and then trying to commit those lists to memory, Wilkins attempted to show how such a scheme could be used to order all human knowledge. We will see later (Chapter 9), how starting with a single universal measure (of length) it is possible to construct a self-contained system of units which is the basis of modern science. If you like, how a truly universal language may be constructed from a limited number of primitive semantic primes, or base units.

John Wilkins did not so much wish to discover the language used by Adam in the Garden of Eden; he wished to be the new Adam, by turning the old mystical speculation of universal languages on its head. As he wrote in the Introduction to his Essay of 1668, "*This design would likewise contribute much to the clearing of some of our modern differences in Religion, by unmasking many wild errors, that shelter themselves under the disguise of affected phrases; which being Philosophically unfolded, and rendered according to the genuine and natural importance of Words, will appear to be inconsistencies and contradictions.*" To fulfill this promise of reshaping language, of creating a tool for linguistic analysis and of providing a means of standardising religious understanding, it was not enough simply to invent real characters for this new language; it was necessary to develop a criterion that would govern the primitive features that would compose these characters. In order to design characters that directly denote concepts and ideas, if not the things themselves that these concepts reflect (this was the problem which Adam was faced with when he was commanded by God, to name Creation), two conditions must be fulfilled: (1) the identification of the true primitive notions or semantic primes and (2) the organization of these primitives into a system which represents the organization of the model of the content. Thus, such a language is termed *a priori*. And the formulation of such a language requires a grammar of ideas that is independent of any natural language.

John Wilkins' ideas were widely circulated among the *savants* of his period. Unfortunately, his ideas were met mostly with derision, not only among ordinary people but also among fellow savants as being brilliant, but incomprehensible. The *Ballad of Gresham College* is a satirical ode on the Royal Society (originally called Gresham College) and refers directly to Wilkins' project,

> *A Doctor counted very able*
> *Designes that all Mankynd converse shall,*
> *Spite o' th' confusion made att Babell,*
> *By Character call'd Universall.*
> *How long this character will be learning,*
> *That truly passeth my discerning.*

Science is the reduction of an extraordinary and bewildering diversity of events and observations into a manageable uniformity within one of a number of possible systems of symbols, quantities, and units. Similarly, technology is the art of using those systems of quantities, units or symbols so as to be able to predict, control and organise events. The scientist always views things through the medium of a system of symbols, quantities, and units, and technology is the handling of information and materials in ways that have been predicted by the systems of symbols and units.

To many this may seem like magic and, unfortunately, the more isolated the scientist becomes from the general public, the more priest-like the scientist appears. But popular or not, communicative or not, science does evolve and impacts everyone. I am attempting to outline here how the modern language of science grew out of magic and the search for a mythical language of power that was used by God to create the Cosmos, and us. Indeed, it was not so long ago that the original search for the proto-language spoken in Eden was not so much abandoned, but subsumed into the search for a universal philosophical language which would better allow man to understand who he is, where he is, and why he is here. This change from the search for the language with which God created the physical world, to the creation of the language of science, probably arose because the later language actually worked at allowing man to speculate about Nature—it produced results. It was seen to be a language of authority, and not just a language of curiosity. The proto-language might have allowed God to create the Heavens and the Earth, but the philosophical languages allowed man to understand the Heavens and the Earth, and permitted him to dream that science might even allow him to one day be able to control Nature.

Science works. But we scientists have still not completely shaken off the aura of the magician or *magus* (whether we know that or not). As we are only too aware, all technologies are increasing in performance at an alarming rate; electronics and computers are becoming smaller, faster, and cheaper. But modern science and technology, and hence today's scientist and the technician, have left the ordinary man far behind; and in the future, technology and magic will, perhaps again, become indistinguishable. But then this was the case in the 17th century, so there is nothing new there.

6.1   FURTHER READING

For all aspects of the rise of science from an occult and magical background, the works of Frances Amelia Yates (1899–1981) are all worth reading. She was an English historian who focused on the study of the Renaissance, and also wrote books on the subject of the history of esotericism. In 1964, Yates published *Giordano Bruno and the Hermetic Tradition*, an examination of Bruno, which came to be seen as her most significant publication. In this book, she emphasised the role of Hermeticism in Bruno's works, and the role that magic and mysticism played in Renaissance thinking. She wrote extensively on the occult or Neoplatonic philosophies of the Renaissance. Her books *The Occult Philosophy in the Elizabethan Age* (1979), *The Art of Memory* (1966), and *The Rosicrucian Enlightenment* (1972) are major contributions, where the author deals with the supposed remoteness and inaccessibility of studies of magic and of the Hermetic arts. These volumes are available from Routledge Classics, Oxford, an imprint of Taylor & Francis.

A truly remarkable history of languages, especially the more esoteric aspects of that history: *The Search for the Perfect Language* (1995); Umberto Eco; Great Britain, Blackwell Publishers Ltd.

CHAPTER 7

The Laws of Nature

Nature and Nature's laws lay hid in night: God said, "Let Newton be!" and all was light.

Alexander Pope: epitaph for Sir Isaac Newton

A moment's thought will demonstrate that science may be described as the quantitative study of the complex, coupled relationships that may or may not exist between observed events. Any phenomenon that is susceptible to investigation, that can be measured, that can be weighed, that can be numbered, and that can be expressed mathematically, the readings on laboratory dials, the clicks coming from a counter, or detector can all be considered as part of the enterprise of science. On the other hand, there is no room in the scientific worldview for the inexact, uncontingent, immeasurable, imponderable, or undefined. A process that can be repeated time after time, a system that can be reproduced and analyzed, these are the concepts which go to make up science, and not the individual, the unique, the elusive thing, or phenomenon that can never occur a second time.

Our increasing understanding of ourselves and of the world within which we live comes from the myriad measurements scientists and technicians make each day. These measurements drive the evolution of our society. We realized in the 17th century that by studying and using the newly discovered Laws of Nature to make predictions about future events we no longer needed magicians or Shamans whose predictions about future events were correct only on a statistical basis. This observation was a significant advance for mankind. Indeed, the history of science is an essential part of our political and economic freedom; it could be said to be the Palladium of our freedoms. One way of thinking about our present democracy is as an expanding mass of conflicting interests, which through the action of a solvent such as modern capitalism, spiked with a fascination for trivia, such as are readily available on the Internet, becomes resolved into what is, in essence, a thin vapor. That is, a dilute or rarefied, ideal gas of non-interacting particles that lose collective internal energy in proportion to the perfection of its aspiration. Like a perfect gas expanding into a vacuum… into nothing, losing all coherence and long-range structure. By using the prism of the scientific worldview to keep our views of how we came to be who we are in perspective, we might even be able to preserve our freedoms.

When we know something of the origins and evolution of our assumed knowledge, or understanding of the natural world, we are delivered from the thrall of preconceived opinions and the foolish, fabulous ideas into which man is all too willing to fall, and into which one may fall without even realizing it. In addition, we better understand the limited value, the limited shelf-life of all our hypotheses about the unfolding of the visible universe, and on a smaller scale, about the evolution of our own society and of our own lives. When we study the history of science, we see how misunderstandings in science have arisen, and how they were resolved. And we are able to put the achievements of our own period in a more appropriate perspective; a more appropriate historical perspective.

A study of history and particularly the history of science tells us how well we have been thinking, and whether what we have been thinking about is relevant and useful. And among other things, a study of the general scope of historical development affords the scrutiny of evidence, and the capacity to decide which particular version of an event seems most credible. It also allows one to observe the strange, almost unfathomable, metamorphosis that occurs in the interpretation and hallowing of a sacred text; invoked as if it were supernaturally ordained, and hence not available for contested examination and interpretation. That is, we may investigate the origins of the dogmas of science; it allows us to understand Nature and to live with less anxiety with the more violent aspects of the natural world (see Chapter 4).

The essential and all-important characteristic of science is that it is predictive. Science follows some established order; an order that was thought by our forebears, even as recently as the late 19th century, to be divinely inspired. Today, however, we hold that phenomena arise because of a set of transcendent fundamental laws, and the interaction between a set of unchanging forces that may be characterized by a set of inviolable constants of Nature, for example, the mass and charge of the electron, m_e and e, respectively. But where did this idea of a divine legislator for the Universe come from? If we can say that the natural world has arisen from, and is maintained by a set of fundamental laws what is the similitude between these observed Laws of Nature and the laws promulgated by national parliaments? Where did the observed Laws of Nature come from?

7.1   THE COMPLEX RELATIONSHIP BETWEEN ASTROLOGY AND ASTRONOMY

In earlier parts of this volume, I made much of the complex relationship between a magical way of looking at Nature, and a more rationalist or scientific way of looking at Nature, that is, the close relationship of natural philosophy, the Hermetic arts, and modern science. Nowhere is this complexity better seen than in a comparison of astrology and astronomy. In addition, in Chapter 5, we saw how the creation of complex, self-contained system such as the *I Ching* and the Houses of the Zodiac formed a system of study permitting a range of correspondences with the observed natural world to be constructed and maintained. If then one projects these theoretical, mystical interconnections onto other observations of Nature, one does have a system of sufficient flexibility to explain some aspects of Nature. But of course, this is a mystical theoretical model of Nature; one could say a mythological model. But this is where it all began. Ancient civilizations, such as the Sumerians, were famous for their ability as both astronomers and as astrologers. You could not be an astrologer if you did not know something of the slow, reproducible, sacred dances of the planets and the stars. Consequently, we first need to consider the origins of the words *astrology* and *astronomy*.

Table 7.1 gives a list of some alchemical and astrological/astronomical symbols commonly used by *savants* in the pre-scientific age (during and before the late 17th century). As can be seen (compare with Figure 3.1), there is a clear mixing of the symbols; the symbols used in alchemy also represent the metals that are associated with the seven planets that also give us our days of the week. The use of these symbols descends from ancient Greco-Roman astronomy/astrology, although their current shapes are a development of the 16th century. The symbols of Venus and Mars are also used to represent the female and the male in science, following a convention introduced by Linnaeus (see Chapter 13) in the 1750s. Even today, it is often difficult to separate the scientific and the magical description of Nature.

Table 7.1: A table of symbols for celestial bodies (astrological and astronomical symbols) and chemical elements (alchemical symbols)

Important Alchemical and Astrological Symbols		
According to the Swiss alchemist and chemist Paracelsus (1493–1541), the three primes or tria prima—of which material substances are composed are mercury, salt, and sulphur. Paracelsus reasoned that Aristotle's four element theory appeared in all bodies as three principles. He saw these principles as fundamental, and justified them by recourse to the description of how wood burns. Mercury included the cohesive principle, so that when it left as smoke the initially solid wood fell apart. Smoke described the volatility (the mercurial principle), the heat-giving flames described flammability (sulphur), and the remnant ash described solidity (salt).		
Mercury (or mind)	■	This is also the symbol for the planet Mercury
Salt (base matter or body)	■	
Sulphur (or the soul)	■	

... continued on following page

Western alchemy makes use of the Hermetic elements. These are the four classical elements of Aristotle: air, earth, fire, and water

Air	■	The properties of the four classical elements are first discussed by the Islamic scholar Abū Mūsā Jābir ibn Hayyān (c.721–c.815). He has been widely described as the father, or the founder of early chemistry, inventing many of the basic processes and equipment still used by chemists today.
Earth	■	
Fire	■	
Water	■	

Seven metals are associated with the seven planets, which also give us our seven days of the week, and seven major deities, all figuring heavily in alchemical symbolism. Although the metals occasionally have a glyph of their own, the planet's symbol is most often used, and the symbolic and mythological septenary is consistent with Western astrology. The planetary symbolism is limited to the seven wandering stars visible to the naked eyes of ancient astronomers, the extra Saturnian planets. Uranus and Neptune are not included, as they were identified as planets only in the late 18th and early 19th centuries, respectively.

Lead dominated by Saturn	♄	Also the symbol for the planet Saturn. Saturday is the day of Saturn or Kronos—*dies Saturni.*
Tin dominated by Jupiter	♃	Also the symbol for the planet Jupiter. Thursday is the day of Zeus or Jupiter—*dies Iovis.*
Iron dominated by Mars	■	Also the symbol for the planet Mars, and for the male. Tuesday is the day of Mars—*dies Martis.*
Gold dominated by Sol (the Sun)	■	Also the symbol for the Sun. Sunday is the day of the Sun—*dies Solis.*
Copper dominated by Venus	■	Also the symbol for the planet Venus, and for the female. Friday is the day of Venus or Aphrodite—*dies Veneris.*
Mercury (quicksilver) dominated by Mercury	■	Also the symbol for the planet Mercury. It is also used as a unisex symbol since the intersex Hermaphroditus was a child of Hermes and Aphrodite (Mercury and Venus). Wednesday is the day or Mercurius or Hermes—*dies Mercurii.*
Silver dominated by Luna (the Moon)	■	Also the symbol for the Moon. Monday is the day of the Moon—*dies Lunae.*

It was the Austrian-American, Marxist historian and sociologist Edgar Zilsel (1891–1944) who pointed out that the compound word, astronomy could not have been formed and used had there not been, at that time a tacit recognition of the quasi-juridical nature of the laws which control the motions of the heavenly bodies. That is, that there was a celestial law-giver who legislated for the Universe. [1]

The *de Legibus* (*On the laws*) is a dialogue by Marcus Tullius Cicero (106–43 BCE) composed during the last years of the Roman Republic. Cicero wrote this work as a fictionalized dialogue between himself, his brother Quintus, and their mutual friend Titus Pomponius Atticus. The dialogue begins with the trio taking a leisurely stroll through Cicero's estate at Arpinum; they begin to discuss how laws should be made, and how they should be maintained. Cicero uses this text for expounding on his theories of natural laws of harmony among the social classes. But what Cicero also included was the comment, "*The universe obeys god, seas and land obey the universe, and human life is subject to the decrees of the Supreme Law.*" Cicero's comment was certainly a Taoist view of the nature of all things, but it was a view that demonstrated a separation between divine laws (Laws of Nature) and the laws of men. Yet, in his de *Natura Deorum* (*On the Nature of Gods*), Cicero tells us that gods and men are influenced by the same laws, so we see there is an indication that there were laws of Nature which bind us all.

The words astrology and astronomy were at first synonymous, and the later was familiar to Aristophanes in the 5th century BCE (*Clouds*, lines 194 and 201). Subsequent usage seemed to follow the personal preference of individual authors. Plato wished to settle on the word astrology for all investigations of the heavens, but astrology was already beginning to acquire the magical significance of astro-mancy. In the astrological literature of late antiquity, we sometimes encounter a mixing of terms, which today we take a great deal of care to separate; for example, "Laws of Nature" are mentioned in the context of a magical interpretation of phenomena. The astrologer Vettius Valens (120–175), while discussing an astrological predetermination (submission to fate), speaks of the legislation of Nature, of fate and of the stars.

Vettius Valens' surviving texts are particularly interesting, because he cites the views of a number of earlier authors and authorities who would otherwise be unknown. Although the astronomer, mathematician, and astrologer Ptolemy of Alexandria (90–168), and author of *Tetrabiblos* (the most influential astrological text we possess), was generally regarded as the colossus of Hellenistic astrology and astronomy for many centuries following his death, it is likely that the practical details of the astrology of the period resemble the methods elaborated in Valens' Anthology. Modern scholars tend to compare and contrast the two men since both were roughly contemporary and both lived in Alexandria. Yet Valens' work elaborated the more practical techniques that arose from ancient tradition, while Ptolemy was more of a "modern" scientist, and tended to focus on creating a theoretically consistent model based on his Aristotelian interpretation of the Cosmos. Ptolemy's model of the Cosmos persisted until the early 17th century.

Deciding that the traditional Pagan religion (with all those sexually driven anthropomorphic gods and goddesses) was useless, Valens found in fate a substitute religion. For him, absolute pre-destination gave emotional satisfaction and aroused an almost mystical feeling of oneness with the Cosmos. Knowing that everything was already predetermined, apparently gave one a sense of freedom from anxiety (*ataraxia* or "unperturbedness") and a sense of salvation. With such a view of Nature, we are not far from the *Consolation of Philosophy* by the late-Roman writer Boethius (480–524/525).

In the 5th century AD, Latin encyclopaedias written for monks explain astronomy as, literally, the "science dealing with the laws of the stars," that is, *lex astrorum*.[13] But these sources are not fully explained, and the significance might be that of the laws which the stars give to every man in fixing his fate, rather than that of the laws which the stars themselves had to obey in their eternal motions. So, we are no further forward.

There is no clear distinction between astrology and astronomy until we get to the European Enlightenment, with even Isaac Newton being both an astrologer and an astronomer. Indeed, the idea of *lex astrorum* suggests that gravity is not only what keeps the stars in their courses, but also what carries astrological "influence." Perhaps it was this later function that started Newton on his great quest for gravity. In short, there is no simple explanation or authority on the difference between astronomy and astrology other than one of personal belief in the influence of the stars on our fate, a fact that is emphasized when we consider how easily we make a Freudian *lapsus* when using the two words.

7.2   THE SEARCH FOR THE DIVINE LAWGIVER

It is quite difficult to locate an exact moment when natural philosophers, *savants*, or magicians started using the term Law of Nature for a law or laws derived from observations of Nature, and which is considered to be inviolable; that is, a law which has absolute authority over all of us, and over our society. Archimedes (Figure 7.1) was probably the first to expound a Law of Nature, but he would probably have regarded what we call the law of the lever, as a principle rather than an inviolable law. But by the mid-18th century, the term Law of Nature was being widely used, certainly as a result of the propagation of the Newtonian synthesis of mathematics, mechanics, and optics, although how many of those who used the term had much of an inkling of what it might mean is a moot point. Of course, there are the majestic lines of hymn, number 535 from the English Hymnal of 1796,

> *Praise the Lord, for he hath spoken.*
> *Worlds' his mighty voice obeyed;*
> *Laws that never shall be broken,*
> *For their guidance he hath made.*

---

[13]   See Cassiodorus, Inst. 2, 7 and Isiodorus (Isidore of Seville), Diff. 2, 152.

It is not entirely clear, however, if the author is here writing about God's law as being a set of rules as given in the Bible, or a set of more fundamental rules stating how the Cosmos itself was to function. Such was the prestige of Newton in the late 18th century that this verse could just as well apply to the demi-god Newton, who had identified and presented to man a set of laws, or principles that he said were inviolable.

Figure 7.1: The Fields Medal. This medal is awarded to those who achieve significant advances in mathematics (there being no Nobel Prize for mathematics), and it carries a portrait of Archimedes (c.287–c.212 BCE), as identified by the Greet text. The Greek natural philosopher was well ahead of his time in using the modern scientific techniques of observation, conjecture, and further confirmatory observations (and experiment) to understand the phenomena he saw around him. *On the Equilibrium of Planes* is a treatise in two volumes by Archimedes. The first book establishes the law of the lever, and locates the center of gravity of the triangle and the trapezoid. According to Pappus of Alexandria, Archimedes' work on levers caused him to remark: "*Give me a place to stand on, and I will move the Earth.*" The second book, which contains ten propositions, examines the centers of gravity of parabolic segments. The Latin phrase states: *Transire suum pectus mundoque potiri* (*Rise above oneself and grasp the world*).

It is almost certain that the concept of a celestial lawgiver legislating for non-human, natural phenomena goes back to the Ancient Sumerians. The Sun god, Marduk, was raised to central pre-eminence in Babylonian mythology about the same time that the sixth king of the First Dynasty of Babylon, Hammurabi (c.1810 BCE–c.1750 BCE), codified his society's laws. We read how Marduk is he who prescribes the laws for the other gods, and it is he who fixes their bounds. Marduk is the lawgiver to the stars. It is he "*who prescribes the laws for the lesser star-gods, Anu, Enlil and Ea and who fixes their bounds.*" Marduk it is who "*maintains the stars in their paths*" by giving "*commands*" and "*decrees*'" (from the Later Babylonian Creation Poem as given in Joseph Needham in *Science and Civilization in China* Volume II, P.533). Similar ideas of a supreme law-giving god may be found in Hindu literature; see the *Rig Veda* X, 121.

Today, we know that it is the mass of the planetary and stellar bodies interacting with each other through the medium of gravity, which holds the stars in their courses; however, this idea of Isaac Newton is barely 300 years old, and received its latest refinement by Albert Einstein only a century ago. And although the ideas of Newton and Einstein are accepted by modern scientists as dogma, they are not widely understood. However, the concept of a primal lawgiving sky-god is still very much accepted by, perhaps, the majority of humanity.

At an earlier period of scientific development and investigation, the pre-Socratic philosophers spoke about "*necessity in Nature*" but not about the "*laws of Nature.*" For example, Heraclitus (c.500 BCE) tells us that "*The Sun will transgress his measures, otherwise the Erinyes, the bailiffs of Dike* (Goddess of Justice) *will find him out.*" Anaximander (c.560 BCE) also speaks of the forces of Nature paying fines and penalties to each other for slights and transgressions. But then is this not what the stories of Greek Mythology are really implying; that behind the lusty gods, goddesses, nymphs, and heroes whose stories are intended to instruct the unsophisticated, there was a complex philosophical picture about the nature of divine and human transgressions and actions. The Roman Stoics maintained, as did Zeno of Citium and Diogenes that Zeus, being immanent in the world was nothing other than universal law, an intelligent presence, or *logos* behind Nature—as with many ideas about the nature of the Monotheist god.

Aristotle makes a separation between "*positive law*" which is obeyed by society, and "*natural law.*" In the *Nicomachean Ethics* (V, vii) we read "*Some people think that all rules of justice are merely conventional, because whereas* [a law of] *of Nature is immutable and has the same validity everywhere, as fire burns both here and in Persia, rules of justice are seen to vary.*" Plato does use the phrase *law of Nature* in the *Timaeus*, but, unfortunately, he did not discuss the subject. It is the Stoics, particularly the domineering, law-conscious Roman stoics who developed the idea of a set of supreme natural laws common to all men, irrespective of their national or cultural heritage. One can see that just as the Babylonian idea of Laws of Nature grew out of a centralized, absolutist oriental monarchy or authority, so in the time of the Roman stoics, living within the world empire of Rome with its greatly increased centralization of power and of authority, it would be natural to view the Universe as a great empire ruled by a divine Logos, or intelligence. A supreme intelligence, which maintained the stars in their courses, and ruled the destinies of all men and of all empires.

It is from the poet Ovid (43 BCE–17 AD) that we find the clearest statement of the belief in the existence of laws in the non-human world (in *Pythagoras* from *Metamorphoses* XV, 17). "*What shakes the earth; what law the stars keep their courses under, and what so ever thing is hid from common sense;*" that is, Pythagoras knew the laws according to which the stars move. Ovid does not hesitate to use the word *lex* (law) for stellar and planetary motions. In the *Tristia*, Ovid describes a supposed friend's faithless behavior as being so appalling as to make rivers flow uphill, the Sun go backward, and all things proceed reversing Nature's laws.

Judaism, Christianity, and Islam are, of course, built on the idea of a single divine lawgiver. Perhaps it was during the Babylonian Captivity that the Jewish people came to adopt the idea of a single transcendent god. Certainly, it is with the Hebrew Bible that we first begin to glimpse a celestial lawgiver who influences both Nature and human society. *"The Lord gave his decree to the sea, that the waters should not pass his commandment"* (Psalm 104) and *"He hath made them fast for ever and ever, he hath given them a law which shall not be broken"* (Psalm 148). The problem with the monotheist view of natural laws, however, was that it quickly became identified with morality; human morality, particularly the do's and don'ts of sex, as Saint Paul and Saint Augustine of Hippo inform us. Yet, even as late as the 4th century AD, the Christian apologist Arnobius of Sicca (died about 330) could argue that Christianity was not such a bad religion after all; as the adoption of Christianity by the Roman Empire had not changed the way the natural world worked. After all, the Sun still rose and the Moon still followed its traditional cycles. That is, that the Laws of Nature are implicit. The rotation of the stellar firmament, the cycles of the seasons had not altered with Constantine's Edict of Milan of 313. Whatever was driving the Universe had little to do with the Christian religion; replacing Jupiter or Jove by God, Yahweh, or Allah in your affections did not change the visible world, it merely influenced your private life.

7.3   A VERY DIFFERENT POINT OF VIEW

What of the idea of a celestial law-giver in the Orient? Following Needham (*Science and Civilization in China*, Volume II, P.554*ff*) we only need look at the *Nei Ching* which contains conversations between Chi Ni Tzu and Kou Chnen, the King of Yŭeh in the late 4th century BCE. The king asks the sage about the origins of natural phenomena (he has already asked him about the forces that rule human society). *"There are the Yin and the Yang. All things have their chi-kang* [that is, their fixed position and motions with regard to other things]. (This *chi-kang* is what Needham translates as "Laws of Nature.") *The Sun, Moon and Stars signify punishment or virtue, and their change indicates fortune and misfortune. Metal, wood, water, fire and earth* (the five elements of Classical Chinese thought; slightly different from the European quartet.) *conquer each other successively; the Moon waxes and wanes completely. Yet these normal changes have no ruler or governor. If you follow it* [heaven's way] *virtue will be attained; if you violate it there will be misfortune."*

The Ancient Chinese viewed Nature as a great net, or vast pattern. There is a web of relationships throughout the Universe, the nodes of which are things and events. There is no ruler or governor. Nobody wove this great net; it is eternal, like the quantum mechanical view of the vacuum, but if you interfere with the texture of the net, you do so at your peril. The Ancient Chinese did not follow the Roman stoic's love for celestial law-givers and law-enforcers. The Ancient Chinese did not need the sense of security coming from the creation of an all-powerful, male deity who lived in the sky, who had a long white beard (a sign of wisdom according to Gnostic creation myths), and who would tell us all what to do and what to think; and of equal importance, what not to do, and who not to do it with.

The idea that heaven does not command the processes of Nature to follow their regular courses is linked to the belief system and philosophy which we know today as Taoism, where, *wu wei* or non-action, or unforced action is central to the ways of heaven. The Tao of Heaven is a *Ch-hang Tao*, the order of Nature is an unvarying order, as was said by Hsün Chhing in about 240 BCE, but that is not the same as affirming that anyone ordered it to be so. As Confucius (c.551 BCE–479 BCE) says in the *Li Chi* "*The most important thing about* [the ways of Heaven] *is its ceaselessness... Without any action being taken, all things come to their completion; such is the Tao of Heaven.*" There is a denial, if only an implied denial, of any heavenly creation or legislation. The heavens act according to *wu wei*; the Tao produces, feeds and clothes the myriads of things that compose our world, it does not lord it over them, and asks nothing in return.

Back in Europe, it is not until the 17th century that *savants* or natural philosophers began to separate morality from the Laws of Nature, which were thought to be obeyed by animals, humanity... minerals, plants, chemical substances and planets alike. This separation could only have occurred with the advent of the Reformation, and the idea that there existed a "right of rebellion" against a supposedly un-Christian prince or authority. That is, a change in the worldview of European man could not have begun until the absolutism of the pre-Reformation Catholic Church had been challenged and successfully broken by the Protestant Reformation; popes such as Alexander VI, Julius II, and Leo X were absolute "oriental" potentates. If it could be accepted that princes could act contrary to natural law, no matter how well or badly defined was that natural law, then a distinction could be made between natural or universal laws or authority, and man-made laws or authority. And, perhaps, in the case of man-made laws they should be more accurately termed as choices rather than as authority. Before the Reformation, the Christian world was in thrall to the greatest of Scholastic philosophers, Saint Thomas Aquinas. This Dominican friar and teacher envisaged a system of sets of laws: the *lex aeterna*, which governed all things for all time, the *lex naturalis*, which governed all men, and the *lex positiva* created by human legislators (-divina if canon law inspired by the Holy Ghost working through the church, and –humania, or common law laid down by princes and governments).

Remarkably though, Johannes Kepler (1571–1630), who discovered the three empirical laws of planetary motion, one of the first occasions when Laws of Nature, or rules of observation were expressed in mathematical form, never referred to them as laws. Indeed, neither Galileo Galilei nor Nicolaus Copernicus (1473–1543) ever used the expression "Laws of Nature," even though their work is the foundation of modern science. Perhaps, we see here the last vestige of the influence of the Roman Church on natural philosophers and the church's desire not to separate the concepts of universal laws and man-made (that is, church-sanctioned) laws for the governance of the Universe and society, respectively. Even Isaac Newton could not bring himself to totally decouple universal laws and society's laws.

7.4   THAT FEARFUL PERFECTION

But change was coming; the *zeitgeist* was moving in the direction of the creation of a God whose relationship to His creation could be examined. The Catholic heretic, Giordano Bruno, following Nicolas de Cusa (1401–1464), asserted that God was a perfect sphere. That is, the most perfect of solid (Platonic) bodies.[14] Xenophanes of Colophon (c.570–c.475 BCE) was the first Classical philosopher to speak against the anthropomorphic nature of the gods, and spoke instead of a single transcendent god (“*One god, greatest among gods and humans, like mortals neither in form nor in thought*”). This perfect deity was conceived of as being a sphere. It was Plato who had told us that the sphere was the most perfect and uniform of all solid bodies; ideas that carried forward to discussions of the shapes of atoms and molecules in the last century. For some Classical writers it was inconceivable that the transcendent god would not be spherical, because this shape was the best, or least inadequate to represent the Divine, the supernatural. But how did this abstract, geometrical image of God become established in the European mind?

Alain de Lille (c1116/1117–1202/1203) was a French theologian and poet who studied and taught in the schools of Paris where he came under the influence of the philosophers and mystics attached to the Augustinian Abbey of Saint Victor. Alain was also influenced by ideas of materialism, which could have condemned him to the flames; he wrote “*God is an intelligible sphere, whose center is everywhere and whose circumference is nowher*e.” This powerful, fearful image of the Divine sphere quickly became part of the European imagination. In Rabelais we read of, “*that intellectual sphere, whose center is everywhere and whose circumference is nowhere, and which we call God*” (*Pantagruel*). The mediaeval mind believed that God was in each of His creatures, but none of them limited Him, “*The Heavens and the Heavens of the Heavens cannot contain thee*” (1 Kings 8: 27). What better image for the Divine than the sphere?

Nicolas de Cusa wrote in his *De Docta Ignorantia* (*On Learned Ignorance*) that, “*Deus est absolutus;*” no arguments or quibbles here. But he was following Saint Anselm of Canterbury (1033–1109), who gave us the first ontological proof of the existence of God, who said God is, “*id cujus nihil majus cogitari possit*” (*something beyond which nothing greater can be envisaged*). God can never fully be reached by the human intellect. One could say that invoking the geometric metaphor of the sphere that the relationship between our knowledge of God and our view of Nature is the same as that between a polygon made up of many (N) sides and the circumference of a circle. As

---

[14] Interestingly, while Bruno was burned at the stake in Rome for such ideas, Nicolas de Cusa had gone on to become a Cardinal; evidently, the Middle Ages was a more easy-going time for cosmological speculation than the late Renaissance, but that difference was probably due to the intervening Reformation.

N increases, the polygon more closely resembles the circumference of the circle which may contain it, but they can never be commensurate. The circle cannot be squared. God is that circle (one slice through a sphere) whose center is everywhere, but whose circumference is nowhere.

But whatever the difference in the disciplinary nature of the church for those who contemplated the nature of God, between the end of the Middle Ages and the late 16th century, both Giordano Bruno and de Cusa demonstrated that there were no crystalline Ptolymeic Spheres between man and the Empyrean, where God was believed to dwell. There was just an immense, boundless emptiness filled with stars like our Sun. Bruno had finally overturned the Aristotelian model of the Universe by accepting the ideas of Copernicus. Before Copernicus and Bruno, when man looked into the heavens at the stars, he believed that he was looking inward toward God, toward the *premium mobile* as given in the cosmology of Dante, so wonderfully expressed in the *Divine Comedy*. After Bruno, when man looked at the stars, he looked into the depths of empty space, which Blaise Pascal (1623–1662) so eloquently and majestically told us terrified him, "*Le silence Eternel des ces espaces infinis m'effraie*" (*Pensées* 102), Dante thought that space was a cathedral containing God and man. Giordano Bruno, however, showed man that he was alone on a seashore looking out into the unknown. The Divine sphere, which contained all things, was fearful indeed.

Giordano Bruno derived this geometric idea of the nature of God/the Universe from Nicolas de Cusa, who in turn had derived the idea from Alain de Lille. But from where or from whom did Alain de Lille get this idea? Interestingly, it seems as if the idea was derived from a 3rd century AD *Corpus Hermeticum*. That is, the idea of God as an infinite sphere filling the Universe, or if you like, by association, the Universe being an infinite sphere, an idea which made the Universe immense and homogeneous and removed man and his small planet from the central position assigned to them by theologians, came from Gnostic cosmology and the Hermetic writings which supposedly derived from the ancient Hellenistic-Egyptian magical writings of Hermes Trismegistus (see Page 55). That is, they date from Alexandria in Egypt and the 3rd century AD, but were supposedly an ancient tradition disappearing back into the mists of antiquity. In particular, from writings which derived from a secret or sealed, self-contained wisdom (*Hermeticum*) relating to alchemy, magic, and philosophy. Such an evolution of a science-like worldview evolving out of magic is not a unique event; it happened again in the 17th ventury. Such a change from magical to a more science-oriented worldview change can be thought of as the change from the use of a language of curiosity to a language of authority to describe the world within which we find ourselves.

The next time this idea of spheres of infinite diameter, but with no ascertainable circumference, was heard of in the writings of the French mathematician, theologian, and philosopher Blaise Pascal who said that "*Nature is an infinite sphere whose center is everywhere, whose circumference is nowhere.*" So, from Alain de Lille to de Cusa to Pascal, via Bruno, we have replaced God with Nature as being infinite. Not only has man lost his centrality from the Christian Cosmos, but even God seems to have got lost in this process.

The new feature of Bruno's universe came from his blending of several philosophical ideas. The monk from Naples was attracted by the atomic theories of the Classical World, which had themselves been associated with the possibility of a plurality of worlds, formed by different combinations of the eternal, never-resting atoms passing in and out of various combinations. Bruno was also fascinated by the idea of Alain de Lille and de Cusa that the Universe had no center yet was infinitely vast. For Bruno, it was the Copernican system that best suited such an unbounded, infinite space, and also provided a model of planetary systems associated with stars extending away from us in all directions. In the same way that the mediaeval philosophers had developed a theology where God was infinite in all his attributes, Bruno correlated this infinite God with an infinite Universe, a physics of the infinite, which corresponded with a theology of the infinite. Bruno was only saying that the divine attributes of God be given physical meaning, just as Isaac Newton would do in the next century when he reconstructed God's omnipotence in terms of an absolute space-time.

Giordano Bruno had affirmed that the Universe was boundless and homogeneous, and that the same physical or natural laws would operate everywhere in this universe; this is still the standard view (dogma) of the physical sciences. Newton's Universal Law of Gravitation is as valid on Earth as it would be in the Orion Nebulae, and Planck's constant has the same value on Earth as it would have on a planet orbiting a star in a distant galaxy. Although Bruno does not use the phrase "Law of Nature" very often (he knew that the Holy Inquisition had their eyes on him), he did frequently refer to *ratio* (reason). He visualized the phenomena we see around us as a synthesis of freely developing innate forces impelling an eternal growth and change. Bruno spoke of heavenly bodies as *animalia* pursuing their course through infinite space, believing in the Neo-Platonic ideal that both organic and inorganic entities and objects were in some sense animated. The *anima* constituted the *ratio*, or inherent law which, in contradiction to any outward force or constraint is responsible for all phenomena underlying motion. This was a very Taoist view of the Cosmos from a 16th-century Neapolitan Dominican friar. He may not have said it often, but Giordano Bruno said that God was to be found everywhere "... *in inviolabili intermerabilique naturae lege*.." (*in inviolable laws of nature*). This made Bruno a Pantheist as far as the church was concerned, although it did demonstrate that Bruno possessed a modern holistic, Taoist, or organic view of the character of natural phenomena.

It was with the triumph of scientific rationalism of the 19th and 20th centuries that we move to definitively speaking about the Laws of Nature, and the advent of science as a language of authority capable of explaining the world around us, and the entire Cosmos. It could not be otherwise; we had removed God from our lives, and humanity wished to assume the divine mantle by showing that all Nature was subject to something that we had discovered and measured. We knew what was happening everywhere in Creation, because it happened in our laboratories here on Earth, particularly, in the Cavendish Laboratories in Cambridge. Whereas to speak of rules or propositions of Nature would have been humbler, triumphalist scientists, however, wished to say that science (and by implication, the scientist) was omnipotent.

We are now at the position to ask why, after such a long period during which the Laws of Nature were viewed in Europe as a theological commonplace, they did attain such a position of central importance in the society of the late 17th century? For example, Pope's epitaph for Isaac Newton in Westminster Abbey, quoted at the beginning of this chapter could not have been written for an earlier *savant* or natural philosopher. How was it that in the early-modern world, the idea of God's sovereignty over the Cosmos shifted from the exceptions in Nature (comets which so terrified the mediaeval, and not so mediaeval mind) to unvarying, absolute, unbreakable rules? The answer is probably to be found in the political changes that were taking place in the wider society of this time. What was it that could lead men to look to an absolutist centralization of power over the Universe? Almost certainly a slow, but inevitable, seepage into Nature of man's conception of an earthly ruler and his sovereignty. Perhaps with the decline in feudalism, and the rise of the capitalist mercantile state with a single central Royal Authority (Henry VIII or Elizabeth I), coupled with a widespread decline in the power of the aristocracy, and an increasing isolation of the monarch as absolute; best demonstrated by that most absolute of monarchs, Louis XIV of France. Perhaps it is no coincidence that the Cartesian idea of God as the supreme legislator for the Universe developed during the lifetime of Thomas Hobbes (1588–1679), "*Nature (the Art whereby God hath made and governs the World)*" (Introduction to *Leviathan*, 1651). Thus, an idea which originated in early Bronze Age Mesopotamia of absolute oriental despotism, was preserved and evolved over three millennia to awake to new vigour in the world of early-capitalist absolutism.

7.5   FURTHER READING

The Social Origins of Modern Science (Boston Studies in the Philosophy and History of Science) by Edgar Zilsel (2000); Boston, Kluwer Academic Publishers.

The Grand Titration: Science and Society in East and West (1969), Joseph Needham; London, Routledge.

In the sections of this present work, where I discuss Ancient China I, will be making use of the magisterial *Science and Civilization in China* (published 1956, re-published 1975), Joseph Needham; Cambridge, Cambridge University Press. In this chapter, I make reference to Volume II of this multi-volume work: *The History of Scientific Thought*.

CHAPTER 8

Measuring the World

…une entreprise [the Metric System] *dont le résultat doit appartenir un jour au monde entier.*

Charles-Maurice de Talleyrand-Périgord (1754–1838)

One of the biggest changes to affect the lives of Europeans in the 16th century occurred in February 1582, when Pope Gregory XIII reformed the solar calendar. This long-needed change should have been instantly accepted throughout the Christian world, but as the Reformation had already splintered Christendom, various nations adopted the new calendar in a piecemeal manner, based on national politics and religious sentiments, with England not adopting the changes until 1753. Russia only adopted the change in 1917. The new Gregorian calendar, named in honor of Pope Gregory XIII, was introduced because the old Julian calendar, introduced by Julius Caesar more that 16 centuries earlier, had made the solar year slightly too long. With the passage of the centuries, this accumulated additional time had become significant and had caused a drift of the seasons, which given the primordial place of agriculture in European society had lead to serious problems. In the Julian calendar, all years exactly divisible by four were leap years. To remedy the trend in the distortion of the solar calendar arising from the imprecision of the Julian calendar, an Italian *savant* Aloysius Lilius (1510–1576) devised a new calendar with new rules: Every year that is exactly divisible by four is a leap year, except for years that are exactly divisible by 100, but the centurial years that are exactly divisible by 400 are still leap years.

The changes proposed by Lilius corrected the drift in the civil calendar, but it was still necessary to delete ten days to bring the calendar back into synchronization with the seasons. This deletion of ten days lead to considerable consternation in Christendom, as ordinary people believed that the church, and the *savants* and natural philosophers who advised the Church were stealing ten days of their lives.[15]

The 16th century was also notable for the widespread introduction of a new idea to simplify everyday arithmetical operations; something that also impinged upon the lives of nearly everyone. That is, the use of decimal numbers (numbers to the base ten). In 1584, the Flemish engineer and surveyor, Simon Stevin (1548–1620) published a set of tables for the calculation of the amount of

---

[15] And even when the new calendar was finally introduced into Great Britain in 1753 (when because of English procrastination it was now necessary to delete eleven not ten days of the year), there was similar popular anger. These events lead to a distinctly anti-science, or anti-expert, attitude in the UK, which persists to this day.

interest that banks would charge for lending money at various rates over various periods of time. As he was preparing these tables, Stevin realized that decimal numbers would greatly simplify calculations in every area of life. Consequently, in 1585, Stevin published *De Thiende* (*Of Tenths*) in Flemish and *La Disme* (*The Tenth*) in French. These were the first books where the simplicity of decimal numbering was fully explained and demonstrated. Thus, the invention of decimal arithmetic is usually attributed to Simon Stevin, who, in addition to assisting people to understand how much interest they were paying to bankers, also found time to use his skill as a mathematician in the design a more efficient type of windmill, which was able to drain land more effectively, and so permitted the creation of the Netherlands.

However, it was in Italy that the greatest scientific advances were being made in the 17th Century; scientific advances that would have a profound effect upon the science of measurement (metrology). The Italian mathematician, astronomer, and *savant* Galileo Galilei enjoyed daydreaming in church. While attending Mass in the Cathedral of Pisa, he allowed his attention to wander from the Liturgy, and it was while contemplating the swaying motion of the heavy chandeliers suspended by long, fine chains from the high ceiling that he formulated several ideas about the pendulum. Galileo went on to conduct detailed experiments on pendulums, and eventually determined the length of a pendulum swinging through its arc in exactly one second in Pisa. This became known as a seconds' pendulum. It was subsequently shown that the seconds' pendulum varied in length according to where it is on the Earth's surface. For example, at the Equator the seconds' pendulum is 991.00 mm in length, and at 45° north of the Equator it is slightly longer at 993.57 mm, this difference arising because of variation is the local value of gravity. The detailed study of the motion of a pendulum undertaken by Galileo made this humble object, a heavy mass attached to the end of a long piece of string, the world's first precision measuring device. It also lead to ideas of occult magic being attached to the pendulum; primarily, but not wholly by Neo-Platonists; as it revealed characteristics of the invisible, hidden, and Hermetic part of Nature—the mysterious force of gravity.

The pendulum is a simple measuring device. The period, T, in seconds of a pendulum of length, L (in meters) is given by $T = (2\pi/\sqrt{g}) \sqrt{L}$, where g is the local value of the acceleration due to gravity. The value of g varies by a few percent over the surface of the earth, and the pendulum is a sufficiently precise device that it is capable of determining the spatial variation of g (approximately, 9.806 m.s^{-2}). This relatively simple relationship yields the approximation, $T = 2 \sqrt{L}$. Thus, the period of oscillation of a pendulum is independent of the mass of the bob of the pendulum. This surprising finding from the 17th century linked the pendulum, in the imaginations of Occultists and those interested in the Hermetic arts with some, as yet undiscovered level of existence; which could well be the much sought-after link between the physical world and the world of the spirit.

An interesting aside on the presentation of decimal numbers, and one which has still to this day not been resolved, is the nature of the decimal marker. In 1615, John Napier (1550–1617) the

Eighth Laird of Merchiston, Scotland, a mathematician, astronomer, and well-respected occultist and astrologer, used a comma as a decimal marker to separate the whole number part from the decimal number part of numbers in his book of multiplication tables *Rabdologia*. Here, Napier was following an idea put forward by the Frenchman François Viéte (1540–1603) who suggested that the comma be used as a *separatrix* between the whole number and the fraction. Unfortunately, for the scientific community, Napier later changed his mind, and replaced the comma as the decimal marker with the full stop. This change from the comma to the full stop as the decimal marker is still with us today. In the English-speaking world, the decimal marker is the full stop, but in the French-speaking world and in Continental Europe, the decimal marker is the comma. This confusion has become an unresolvable cultural identifier.

We saw in Chapter 3 that in 1668 John Wilkins published *An Essay Toward a Real Character and a Philosophical Language*. Wilkins' long essay included a four and a half page description of a proposed system of measurements based upon his idea for a single "universal measure" that could be used to define length, weight, volume, and money. John Wilkins suggested a decimal system of measurement, with a universal standard of length based on time and derived through the use of a swinging seconds' pendulum, and that this standard length could then be used to define area, volume, and weight using a well-defined volume of pure, distilled rainwater. Wilkins' *Essay* is the first description of a complete system of measurement intended to be used by all nations. Indeed, Wilkins' proposal contained almost all of the essential elements of the Metric System of 1795, which could quite reasonably therefore be said to have originated in England in the 17th Century and not in France during the late 18th century (please note that national chauvinism is particularly strong in this debate).

Following John Wilkins' first description of an international system of measurement, the development of the decimal metric system of measurements was inevitable even though Wilkins himself was not confident of its success. Wilkins wrote about his plans for a universal measure, "*I mention these particulars, not out of any hope or expectation that the World will ever make use of them, but only to show the possibility of reducing all Measures to one determined certainty*." Following the publication of John Wilkins' essay, savants in several countries took up and promoted his ideas. The *zeitgeist* was waiting for this concept of a single system of units or weights and measures (and money) based on a single natural dimension. In 1670, Gabriel Mouton (1618–1694), a French cleric and astronomer, promoted a system of measurement that was to be based upon the physical dimensions of the Earth; rather than a measurement based on the length of a seconds' pendulum, or one or other measurements of a human (all be it, Royal) body.

Gabriel Mouton assumed that the Earth was a perfect sphere, and so a section along a Meridian would be a circle. Mouton proposed that this "great circle" should be divided into ever smaller angles, and that these small angles could be used to define a new system of measurement. What was also proposed was that the division of these angles should be made using decimal arithmetic; that is, division by ten, rather than the old Babylonian sub-divisions of an angle based on arithmetic to the base 60 (an angle being divided into 60 min and each minute into 60 sec). Mouton suggested that a minute of arc along a Meridian be measured and defined as a unit of distance called a *milliare*; a linear distance that subtended this angle. The Abbé Mouton also suggested dividing the *milliare* into *centuria*, *decuria*, *virga*, *virgula*, *decima*, *centesima*, and *millesima* by successively dividing by factors of ten. In short, Mouton suggested that Simon Stevin's 1585 decimal system of tenths should be used to divide the Earth-based angular units into ever smaller parts.

Interestingly, Abbé Mouton's *milliare* is the modern definition of a Nautical Mile (that is, one minute of arc of Latitude along any Meridian), which given the importance of maritime trade to the world's major powers accounts for the longevity of this system of measurement. Mouton's *virga* would be one thousandth of a minute of arc, which would be about 1.11 m in today's Metric System. In the same way that Wilkins suggested using a seconds' pendulum beating 1 second, and hence having a length of about one meter, to define the universal measure of length, Mouton suggested using a shorter pendulum to measure the *virgula* (a tenth of a *virga*), or 0.111 m. A pendulum of this length would be beating or oscillating every 0.66 sec.

However, the theoretical models of Wilkins and Mouton demonstrated that to make progress in defining more precise units of measurement; that is, establishing a universal system of measurement to assist in the progress of science and society, one needed accurate measurements of our planet. Was the Earth a sphere, or not; and if not, what was the eccentricity of the planet? One of the first surveyors who undertook the task of precisely determining the curvature of the Earth was the founder of a dynasty of French mathematicians and astronomers, Jacques Cassini (1677–1756) who made measurements of the Earth based on minutes of arc. With his son, César-François (1714–1784), Jacques Cassini surveyed a portion of the Arc of the Meridian from Dunkirk on the coast of northern France to Barcelona in Spain; this is a line from Pole to Pole passing through Dunkirk and Barcelona. This particular Arc of the Meridian also passes through Paris (the Paris Observatory was built on the Meridian line in Paris) and is therefore called the Paris Meridian, and it would be surveyed many times over the following century. These repeated measurements of ever-increasing precision would finally yield the first standard meter, which is still preserved in Paris (the task would be completed by Jacques' grandson Jean-Dominique, 1748–1845). [16]

---

[16] The Paris Meridian is not our present line of zero Latitude. The Prime Meridian runs through the Greenwich Royal Observatory, and so it is at Greenwich and not Paris that the world is bisected and the zero of Latitude established. This move to a Greenwich-based view of the world was decided by an international conference in 1884 (International Meridian Conference 1884, Washington DC,); and, as one can imagine, this move pleased the British, then at the zenith of Empire, and greatly displeased the French. Modern satellite measurements of the geographical coordinated of the Paris Observatory show how much the Meridian had been shifted by global geopolitics; the Paris Observatory is at 48°50'0"N 2°21'14.025"E. The Greenwich Royal Observatory is at 51°28'40.12"N 0°00'0.5.31"W. By using the Abbé Mouton's system of units (given above) one can see how little the Meridian changed geographically, but the political repercussions where tremendous. The French did not begin to accept the 1884 change until after World War II.

Between 1735 and 1737, the explorer, geographer, and mathematician Charles-Marie de La Condamine (1701–1774), the astronomer Louis Godin (1701–1780), and the naval architect, mathematician, astronomer, and geodesist Pierre Bouguer (1698–1758) measured an Arc of a Meridian in Peru where they also made equatorial measurements of the local value of the acceleration due to gravity (*g*). In addition, they returned to Europe with the first detailed map of the Amazon basin. And between 1739 and 1740, the astronomer Nicolas Louis de Lacaille (1713–1762) who had started his career in the church, but then moved to astronomy, together with Jacques Cassini again measured the Dunkirk-Barcelona Meridian. The northern and southern ends of the surveyed meridian were the belfry in the center of Dunkirk and the fortress of Montjuïc in Barcelona, respectively. Apart from defining the dimension of the universal measure or meter, these early surveyors refined the value of the Earth's radius, and definitively established that the shape of the Earth is oblate or slightly flattened near the North and South Poles, which had been predicted by Isaac Newton. This observation lead to the Enlightenment cult of Newton as Universal Genius (see Figure 8.1).

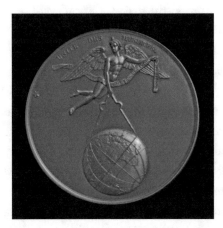

Figure 8.1: A medallion struck in Paris in 1840 to mark the final introduction of the Metric System in France, and to act as a souvenir to posterity of the manner in which the meter was determined in 1799 by a measurement of distance; the allegorical figure is measuring one quadrant of our planet. The reverse side of this medallion bears Condorcet's famous rallying call for the Metric System, A TOUS LES TEMPS: A TOUS LES PEUPLES. This image is reproduced with the permission of the BIPM (https://www.bipm.org/en/about-us/), which retains full international copyright.

While the French were surveying Latin America and Europe, the British finally got around to accepting the Gregorian calendar. Part of the reason given for the UK's decision to finally adopt the Gregorian reforms to the Julian Calendar, and the necessary change in the date of the beginning of the new year, was the difficulty of calculating interest on loans which were outstanding. It was noted, "…[the comparison of the date in England compared with the date on the Continent was] *attended with divers inconveniences, not only as it differs from the usage of neighboring nations, but also from the legal method of computation in Scotland, and from the common usage throughout the whole kingdom, and thereby frequent mistakes are occasioned in the dates of deeds and other writings, and disputes arise therefrom…*" And in 1753, the New Year in Britain actually began on January 1st rather than March 25th, to bring it into line with the rest of Europe because Great Britain (and the British Colonies, including America) began to use the Gregorian calendar (New Style or N.S.) rather than the Julian calendar (Old Style or O.S.).[17]

As the 18th century drew to its close, two political events occurred which would have a profound influence upon the nature of the various systems of weights and measures still used throughout the world today. In North America, the British colonists decided that they did not need to pay the government in London for protection against the French, Spaniards, and Native Peoples. The rebellion of these colonists was successful, and by 1791 a new nation was born, which was using the same system of weights and measures as they had inherited from England. But would they wish to continue using this system? In Europe, the major political event of this period was the bankruptcy of the Kingdom of France, and the collapse of the nation into the French Revolution of 1789.

History tells us that systems of weights and measures are mostly reformed or changed during, or just after, major political upheavals. Well before the Revolution of 1789, everyone in France knew that the systems of weights and measures in France were tangled, convoluted, complex and an invitation to fraud, but no one thought that there was much to be gained by reform. Why reform the system, it was only the ordinary people who suffered adversely from the complexity of the various systems of weights and measures, whose local variations were maintained for the personal advantage of the local aristocrats and church leaders. At the time when the French economy was beginning to industrialize in response to similar developments in England, which essentially had a national system of weights and measures by this time, France was unable to compete as it did not have any means of standardising and automating manufacture. In England, factories could out-produce manual manufacture, because there was uniformity of measurement and standardisation of production based on the inch and the pound.

The, as yet unborn, U.S. sent ambassadors to France, which was actively helping them in their struggle against Britain. Thomas Jefferson (1743–1826) served as Ambassador to France where he

---

[17] Any reader of the letters of the Fourth Earl of Chesterfield to his sons will have remarked that the Earl of Chesterfield was always fastidious in noting whether he was using the Old Style notation for dates, or the New Style notation when dating his letters. But then the reason why the Earl of Chesterfield paid so much attention to the change in calendar was because he was responsible for the change. The Calendar Act (also known as Chesterfield's Act) of 1753 made provision to ensure that monthly or yearly payments would not become due until the dates that they originally would have fallen due under the old Julian calendar. For this reason the UK tax year continued to operate on the Julian calendar and begins on the April 5th, which was the Old Style date for the New Style tax year that began on March 25th.

was in regular contact with British and French *savants* as they formed their ideas about new, more natural systems of units of measurement for science and for society. Similarly, Benjamin Franklin (1706–1790) signed an alliance between France and America, but although the primary objective of this alliance was the raising of funds for the war against Britain, Franklin did not see any reason why he should not take the opportunity of this alliance to promote the cause of science. And it was from this exchange of ideas that the political leaders in America began to consider the best system of weights and measures for their young nation, which would ensure that they were able to compete effectively and independently on an international stage, and not be tied in any way to Britain. For example, in 1786, five years before the American Colonies gained their independence Thomas Jefferson proposed that the new nation adopt a decimal system for their currency. The Continental Congress established the silver dollar as the basis for decimal coins, although it was not minted until after independence in 1792.

At the same time, Thomas Jefferson independently proposed a system of weights and measures very similar to the proposed French decimal Metric System. He differed from the French in that he wanted the meter to be defined by the length of a pendulum that beat a second rather than surveying the surface of the Earth. Jefferson rightly reasoned that other countries could then readily duplicate such a standard at any time; thereby laying the foundations for a truly international science. Jefferson did not particularly like the idea that the meter would be based on a series of surveys made on French territory. Unfortunately, it was at this time that detailed measurements showed by how much gravity varied over the surface of the Earth, and the swinging pendulum definition for the universal measure of length was losing widespread support among *savants*.

During his period as Ambassador to France (1785–1789), Jefferson visited London in 1789. While in London, the political situation in France deteriorated, and to avert bankruptcy King Louis XVI convened the *États Généraux* or States-General for the purpose of imposing new taxes on the nation. The *États Généraux* was a meeting of the three "states" or groups of people who were seen as constituting the nation: the First State was the clergy, the Second State was the nobility and the Third State was the *bourgeoisie*. The urban workers and the peasants were rather left out of things.

8.1   DEFINING THE SIZE OF THE WORLD

Shortly after the fall of the Bastille in July 1789, but long before political stability was re-established throughout France, the science commission of the *Académie des sciences* in Paris recommended a measurement of the new standard of length, the meter, based on a detailed survey along the meridian arc extending from Dunkirk to Barcelona, which had already been surveyed and measured by de Lacaille and César-Francois Cassini in 1739. The commission calculated that if they could measure a significant piece of the Meridian, the rest could be estimated. Both ends of the line to be surveyed needed to be at sea level, and as near to the middle of the Pole-to-Equator Quadrant as possible to eliminate errors. Fortunately for them, the only one such meridian on Earth is about a tenth of the distance (about one thousand kilometers) from the Pole to the Equator and it runs through Dunkirk and Barcelona, so most of the distance to be surveyed lay conveniently inside France, a fact that did not escape the more nationalistic attention of observers such as Thomas Jefferson.

Condorcet appreciated the potential for such nationalist views when he wrote *"The Academy has done its best to exclude all arbitrary considerations—indeed, all that might have aroused the suspicion of its having advanced the particular interests of France; in a word, it sought to prepare such a plan that, were its principles alone to come down to posterity, no one could guess the country of its origin."* The Legislative Assembly endorsed the proposal from the *Académie des sciences*, directed that the detailed survey be made as soon as possible, and enacted the necessary legislation on March 26th, 1791.

Although the *Académie des sciences* finally chose that the meter would be exactly a ten millionth of the distance between the North Pole and the Equator, their choice also defined this distance as being precisely 10,000,000 m. Unfortunately, an error was made in the commission's initial estimation, because the wrong value was used in correcting for our planet's oblateness. We now know that this Quadrant of the Earth is actually 10,000,957 m. One should never forget that these *savants* were not only setting out to create what they saw as a new fundamental system of units based on the dimensions of the Earth, but they were also imposing models and views about the character of the Earth. In 1791, a handful of French Enlightenment mathematicians, guided by the writings of Isaac Newton, imposed a definite shape and size to our planet. The Earth shrank, and became precisely known. The *Académie des sciences* presented humanity with a *fait accompli*. The medallion shown in Figure 8.1 was struck to commemorate this standardization of the Earth.

8.2   OTHER SURVEYS

The European Enlightenment had come up with the idea of constructing a new decimal metrology based on a single measurement of length. Such ideas, however, have a long history, and it is to Ancient China that we must turn for the first consistent use of decimal weights and measures; particularly, in the decrees of the first emperor, Chin Shih Huang Ti in 221 BCE.

Also, given the size of China, it is perhaps not surprising that an early effort was also made in fixing terrestrial length measurements in terms of astronomical measurements or observations. It was an early idea of Chinese *savants*, going back before the time of Confucius (551–479 BCE), that the shadow-length of a standard height (an 8-ft gnomon), at the summer solstice increased by 1 inch for every thousand *li* (a length measurement equivalent to 1,500 chi or Chinese feet) north of the Earth's "center," and decreased by the same proportion as one went south. This rule of thumb remained current until the Han Dynasty (205 BCE–220), when detailed surveying of the expanding Chinese Empire showed it to be incorrect. But it was not until the Tang Dynasty (618–907) that a systematic effort was made to determine a range of latitudes. This extensive Tang survey had the objective of correlating the lengths of terrestrial and celestial measurements by finding the number of *li* that corresponded to 1° of polar altitude (that is, terrestrial latitude), thereby fixing the length of the *li* in terms of the Earth's circumference. This Chinese meridian survey takes its place in history between the lines of Eratosthenes (c. 200 BCE), and those of the astronomers of the Caliph, al-Ma'mūm (c. 827), but more than 1,000 years before the French metric survey of the 1790s.

The majority of these Chinese surveying measurements were undertaken between 723 and 726 by the Astronomer-Royal Nankung Yüeh and his assistant, I-Hsing, a Buddhist monk. The survey was carried out at 11 sites along a meridian running from the Great Wall in the north to Indo-China in the south, a distance of 7,973 *li* or about 2,500 km. The main result of this field work was that the difference in shadow length was found to be close to 4 inches for each 1,000 *li* north and south, and that the terrestrial distance corresponding to 1° of polar altitude was calculated to be 351 *li* and 80 *bu* (the *bu* was a measure of between 5–6 *chi*). The imperial surveyors had achieved their goal of defining a terrestrial unit of length, intended for use throughout the empire, in terms of the dimensions of "Heaven and Earth," that is, 1/351 of a degree.

This survey is today practically unknown, yet it represents an outstanding achievement given the spaciousness and amplitude of its plan and organization, and represents one of the earliest uses of advanced mathematics which was needed to compute the final result. These results were known in 18-century Europe, as they were commented upon by Leonard Euler and later by Pierre Simon de Laplace. While the metric survey obtained a routine precision of about 1 part in 10^6 in distance, the much earlier Chinese survey could boast only of a precision of 1 part in 10^3. The Tang value of the *li* gives a modern equivalence of 323 m, but the earlier standard Han *li* is very different at 416 m.

8.3   FURTHER READING

These books provide a readable background to the origins of the modern metric system.

1   *Defining and Measuring Nature: The Make of all Things* (2014); Jeffrey H. Williams; San Rafael, CA, Morgan & Claypool. This work contains an explanation of the recent redefinitions of several of the base units of the modern metric system.

2   *Le nombre et la mesure : Logique des classifications métriques et prémétriques* (1980); Franck Jedrzejewski; Diderot multimédia (in French.)

3   *The Measure of All Things: The Seven-Year Odyssey that Transformed the World* (2002); Ken Alder; London, Abacus, an imprint of Little, Brown Book Group. This work describes in enjoyable details the problems of the two French surveyors who determined the value of the universal measure, or meter in Revolutionary France during the 1790s.

CHAPTER 9

Dividing Apples with Oranges to Make the Language of Science

In questions of science, the authority of a thousand is not worth the humble reasoning of a single individual.

Galileo Galilei (1564–1642)

Having looked at the evolution of the scientists' worldview, that is, how science compartmentalises and classifies the phenomena and things we observe in the world, let us now turn to how the quantitative is introduced into this classification. After all, science would be nothing but magic without a means of quantifying as well as qualifying what we see around us. We have only words and numbers, the universal currency of humanity to interpret, describe, and record the wonders of Nature. Our various vernacular languages have evolved with our on-going, never-ending study of Nature; indeed, one could say that languages have created man, rather than man having created languages. The worldview of modern science has evolved by specialization of the basic language used to describe natural phenomena, but few understand what has become a highly specialized language. It has become a dialect of an elite. An elite as separate from the general mass of society, as any caste of priests.

Previously, we saw something of the ideas of John Wilkins for a universal philosophical language, capable of being understood by all humanity. By the end of the 17th century, such ideas of a rational universal language were very much part of the European *zeitgist*. In 1666, the German polymath Gottfried Wilhelm Leibnitz published his *Dissertatio de arte combinatoria* in which he claimed that a proper or true philosophical language would be able to analyze all possible concepts into their simplest elements, into what Leibnitz termed "*the alphabet of thought*" (see Page X?). In such a philosophical or, as we would say today, scientific language, a proper symbol should indicate the nature of the animal, phenomenon, or whatever it was naming. In other words, it was a language which could define that thing, or that phenomenon by means of that thing's appearance, or that phenomenon's intrinsic properties. Leibnitz's theoretical proposition presupposed that: (i) ideas can be analyzed into primitive notions or components; (ii) ideas can be represented symbolically; and (iii) it is possible to represent the relations between these ideas. Gottfried Leibnitz was writing in a century which had attempted the construction of many universal philosophical languages, and so presupposed that a complete enumeration of human knowledge could be achieved. The question then arises as to how a relatively small number of fundamental primitive components or base units could be manipulated or combined to produce a true universal scientific language capable of describing all Creation.

The philosophical languages of the 17th century attempted to reform natural languages by simplifying the complex, multiple meanings of some words and concepts. Consider an attempt at learning the definitions of all the words in a dictionary, or of attempting to comprehend all aspects of a discipline of science. In the dictionary, you will find every word defined in terms of other words; in the scientific discipline, you will find explanations involving other scientific terms. In your determination to learn the meaning (or meanings) of every word, you may find that you need to consult the definitions of the words employed in the definitions of other words. Indeed, you soon realize that your initial attempt at learning the meaning of each word in the dictionary is futile. In fact, it is a circular task, because the dictionary contains only a closed set of words, finite in number, that enable descriptions of the meanings of each other. If you do not already have in your mind a set of basic words whose meanings you know independently, without the need of words to define them, you will remain forever in a continuous circular loop with your dictionary; and the same goes for seeking to learn a new scientific discipline. For this reason, the philosophical languages of the 17th century did not reform and simplify English, but they did give us the thesaurus.

At the time of the French Revolution, *savants* who were familiar with the ideas of Wilkins assumed that the new fundamental unit of length, the meter, could be used to define all the scientific and technological concepts required by their society. This meter, or universal measure, was to be defined from the dimensions of the Earth, as one ten millionth of a quadrant running from the North Pole to the Equator. This universal unit of measurement may be thought of as a semantic prime of the new language of science. In fact, it is one of the seven base units, or semantic primes of the International System of Units (SI), which is the modern scientific version of the metric system of 1795; see Figure 9.1. [1][18]

Having defined the basic unit of length, l, to define an area, a two-dimensional quantity, you simply multiplied two distances, that is, $l.l = l^2$. Similarly, when we go to three spatial dimensions to define a volume, we write algebraically $l.l.l = l^3$. Then, assuming that the density (that is, the mass of a known volume of a substance) of, for example, pure water is taken to be well defined as one gram for each cubic centimeter, one can define a base unit of mass as the weight of a precisely known volume of pure water. The kilogram was originally defined as the mass of 1,000 cubic centimeters or 1,000 cm^3, that is, one liter.

---

[18] SI stands for *Système international des unités*, the French name for this set of units; in all matters relating to the metric system, French is the only official language. This point is rarely born in mind by English-speaking nations; but as is pointed out in Chapter 14, politics and science do not always sit well together.

Figure 9.1: Medal commemorating the centenary of the Meter Convention of 1875, manufactured by R. Corbin, *Monnaie de Paris*. This face of the medal represents the seven base units of the SI (meter, kilogram, second, ampere, kelvin, mole, and the candela), and how the meter is defined in terms of the wavelength of light (in 1975 this was via the red light from a krypton discharge lamp) rather than by an artifact. This image is reproduced with the permission of the BIPM which retains full international copyright (https://www.bipm.org/en/about-us/) .

But what happens when we wish to consider the combination of the universal measure of length with other quantities which are essential, in even some of the simplest ideas and concepts of technology, for example, how does one introduce time, the basic unit of which is the second into a system of mechanical quantities? [2][19]

The speed, or velocity, of a planet flying through space, or of an ox ploughing a field, is defined in terms of distance and time; yet, how do we combine these two different base units? One might think of these two quantities as being as different as apples and oranges, so how can they be divided or multiplied together when they certainly cannot be added of subtracted? It is possible to mix and manipulate dimensions of distance and time, to divide or multiply meters and seconds, or even furlongs and fortnights. It is the mathematical definition of a unit that allows us to manipulate distance and time, and generate new ideas such as the concept of speed or velocity, and of acceleration which is speed or velocity per unit time. First, consider what we mean by a unit. Any value of a physical quantity, Q, may be expressed as the multiplied product of a unit [Q] and a

[19] The second is a base unit of the SI, and is the oldest measured quantity having been defined about 5,000 years ago by the Ancient Sumerians.

purely numerical factor (that is, a simple number). Written algebraically, we have Q = (a number). [Q], where [Q] is the unit, for example, meters or seconds, and there are a certain number of these meters or seconds; for example, Q_{length} = 10 meters or Q_{time} = 10 seconds.

This convention of expressing a quantity as a unit and a numerical factor is used throughout science and is referred to as quantity calculus. When units are being manipulated, one may only add like terms, as with apples and oranges, but all units may be manipulated algebraically. When a unit is divided by itself (that is, meters/meters or seconds/seconds), the division yields a dimensionless number, which is one (1) and so intrinsically without dimension or unit. When two different units are multiplied or divided, the result is always a new unit, referred to by the combination of the individual units. For instance, in the SI, the unit of velocity is meters per second; that is, meters/seconds or m/s or m.s^{-1}. This new unit is neither length nor is it time, but length divided by time. When dividing length by time, one is only dividing the numerical factors, which appear before the unit. The two original units are distinct, and cannot be divided but are left as a new unit, meters divided by seconds. Length and time are base units, but the new unit of speed, or velocity is said to be a derived unit, and may be deconstructed into base units. Likewise, density is defined as the mass of a known volume of something, or mass per unit volume. This derived unit is composed of two base units, the base unit of mass (kilogram) and the base unit of length (meter), which as we are dealing with a volume is cubed. Again, we have divided two base units together to create something new.

When the metric system was first introduced in April 1795, there were two base units, the meter and the kilogram; the second was already part of the social fabric. As science and technology advanced in the 19th century, the new profession of scientist (first defined in revolutionary France—*scientifique*) understood how the various manifestations of, for example, heat and work were all related to the concept of energy, and how this idea related to the established base units of length, mass and time. In fact, today we have seven base units which may be combined to explain every known scientific phenomenon, and which would be used to comprehend scientific discoveries that have yet to be made. That is, it is through these seven base units that the true universal language, the language of authority that is science is formulated.

By convention, all physical quantities are organized into a system of dimensions. Each of the seven base quantities used in the modern SI is regarded as having its own dimension. The symbols used for the base quantities or base units, and those which are used to denote their dimensions are given in Table 9.1.

All other quantities, all the phenomena known to modern science are derived from these seven base quantities using the well-established equations, or Laws of Nature and are called derived quantities. As outlined above, the dimensions of the derived quantities are written as products of powers of the dimensions of the base quantities using the equations that relate the derived quantities to the base quantities.[20]

Table 9.1: Base quantities and their dimensions, and the base units of the SI				
Base Quantity	Symbol of Base Quantity	Dimensional Symbol*	SI Base Unit	Symbol of SI Base Unit
Length	l	L	Meter	m
Mass	m	M	Kilogram	kg
Time	t	T	Second	s
Electric current	i	I	Ampere	A
Temperature	T	Θ	Kelvin	K
Amount of substance	n	N	Mole	mol
Light intensity	I	J	Candela	cd

* The dimension of a physical quantity does not include magnitude or units. The conventional symbolic representation of the dimension of a base quantity is a single uppercase letter in Roman (upright) sans-serif type (these specifications are part of the dogma of science [1]).

9.1   CREATING EXPRESSIONS IN THE LANGUAGE OF SCIENCE

Dimensional analysis, or the manipulation of quantity calculus is a powerful tool in understanding the properties of physical quantities, independent of the system of units used to measure them. Every physical quantity is some combination of the base units in Table 9.1, for example, speed, which may be measured in meters per second (m/s) or miles per hour (miles/h), has the dimension L/T, or $L.T^{-1}$, and pressure which is a mass pressing down on an area (as in pounds per square inch) is M/L^2 or $M.L^{-2}$. Dimensional symbols and exponents are manipulated using the rules of algebra; for example, the dimension of area is written as L^2, the dimension of velocity as $L.T^{-1}$ (meter per second), the dimension of acceleration (the rate of change of velocity with respect to time) is written as $L.T^{-2}$ (meter per second squared; that is, meter per second per second), and the dimension of density as $M.L^{-3}$ (kilogram per meter cubed).

Dimensional analysis is routinely used to check the plausibility of newly derived equations, the design of experiments, and the results of calculations in engineering and science before money and effort is expended on detailed measurements. In this way, reasonable hypotheses about complex

---

[20] As mentioned above, the dimension of any quantity Q is written in the form of a dimensional product; dimensions of $Q = L^{\alpha} M^{\beta} T^{\gamma} I^{\delta} \Theta^{\varepsilon} N^{\zeta} J^{\eta}$, where the exponents $\alpha, \beta, \gamma, \delta, \varepsilon, \zeta$, and η are generally small whole numbers (integers), they can be positive or negative, or even zero, and are termed dimensional exponents. This expression defines the make of all things [2].

physical situations are examined theoretically, to see if they merit subsequent testing by experiment. And, it is also the means by which one seeks to determine appropriate equivalent values for a quantity in another system of units; for example, how you convert from the value of a quantity in metric units to the equivalent quantity in British customary units; for example, meters/second to miles/hour, or joules (the SI derived unit of energy, symbol J, where J is equivalent to kg.m^2.s^{-2}; that is, L.M^2.T^{-2}) to British Thermal Units or BTU (a customary unit of energy equal to about 1,055 joules. A BTU is approximately the amount of energy needed to heat 1 lb (0.454 kg) of water, which is exactly one tenth of a UK gallon, or about 0.1198 U.S. gallons, from 39°F to 40°F , or 3.8°C to 4.4°C). Thus, dimensional analysis is the means of translating between the various dialects of the single, universal language of science.

Consider the concept of force, something that is done to an object to make it change its speed, or velocity through, for example, acceleration. In the SI, the unit of force is the newton (symbol, N), named after Isaac Newton in recognition of his fundamental work in mechanics. The newton is equal to the force required to accelerate a mass of one kilogram at a rate of one meter per second squared. In dimensional analysis using Newton's famous formula where force (F) is given as being equal to a mass (m) multiplied by acceleration (a), that is F = m.a, multiplying m (kilogram) by an acceleration a (meter/second^2), the dimension of the newton is found to be M.L/T^2 or M.L.T^{-2}, that is, kg.m.s^{-2}. The newton is derived from the base units of mass, length and time, and so could have been derived by the *savants* of the late 18th century.

These principles of dimensional analysis were known to Isaac Newton, who referred to them as the Great Principle of Similitude. The 19th-century French mathematician and Egyptologist Joseph Fourier (1768–1830) made important contributions to dimensional analysis based on the idea that physical laws like Newton's famous law, F = m.a, should be independent of the systems of units employed to measure the physical variables. That is, the Laws of Nature and fundamental equations should be equally valid in the metric system of units as in a non-metric system of units. And when converting between these two systems of units, we need only be cognizant of the mathematical factors needed to convert between the base units to convert the entire quantity from one system to another. Thus, one should take care never to mix systems of units, as the consequences could be disastrous (see Page 97). But there is nothing stopping one defining force in Ancient Egyptian units of measurement; distance would be in terms of the Royal cubit (about 0.525 m), mass would be in *deben* (about 0.015 kg) and time would have been in *unut* (the hour, which is identical with our hour). Fourier showed how each of these base units would need to be converted to SI base units to convert the Ancient Egyptian unit of force to the newton or vice versa.

9.2 DERIVED UNITS

The base quantities of the SI given in Table 9.1 are combined to generate derived units, which are products of powers of base units without any numerical factors (other than 1). Such a set of coherent derived units is, in principle, without limit and they represent the means by which all the phenomena of Nature are described. Table 9.2 lists some examples of derived quantities and how they are represented in the technical literature.

Table 9.2: Derived quantities

Quantity	Derived Quantity	Representation as a Unit
Area	square meter	m^2
Volume	cubic meter	m^3
Velocity or speed	meter per second	$m.s^{-1}$
Acceleration	meter per second squared	$m.s^{-2}$
Density	kilogram per cubic meter	$kg.m^{-3}$
Surface density	kilogram per square meter	$kg.m^{-2}$
Specific volume	cubic meter per kilogram	$m^3.kg^{-1}$

In this way, the semantic primes of the universal language of science, the base units of the SI, are combined to produce a means of describing and quantifying Nature. Some important derived units are given a specific name, usually to honor the scientist most closely associated with that quantity. Some of these named derived units are given in Table 9.3.

Table 9.3: Named derived quantities

Quantity	Derived Unit (Symbol)	Representation as a Unit
Frequency	hertz (Hz)	s^{-1}
Force	newton (N)	$m.kg.s^{-2}$
Pressure	pascal (Pa)	$m^{-1}.kg.s^{-2}$
Energy (or work)	joule (J)	$m^2.kg.s^{-2}$
Power (or light intensity)	watt (W)	$m^2.kg.s^{-3}$

Of particular interest are two derived quantities related to angles. The plane angle is a two-dimensional quantity defined by two lines, and the solid angle (steradian) is a three-dimensional quantity defined by a cone with a certain cross-sectional area. When expressed in terms of base units of the SI, these two angles are: meter/meter and (meter squared)/(meter squared), respectively; consequently, they are dimensionless, as $m/m = 1 = m^2/m^2$. The fact that quantities related to angles in the SI are essentially invisible, as far as the unit is concerned, needs to be remembered as we see in Table 9.4, which contains more derived quantities.

Table 9.4: Further derived quantities		
Quantity	**Derived Unit (Symbol)**	**Fuller Representation**
Moment of force	newton meter (N.m)	$m^2.kg.s^{-2}$
Viscosity	pascal second (Pa.s)	$m^{-1}.kg.s^{-1}$
Surface tension	newton per meter (N/m)	$kg.s^{-2}$
Angular velocity or torque	radian per second (rad/s)	$(m/m)s^{-1} = s^{-1}$
Heat density	watt per square meter (W/m²)	$kg.s^{-3}$
Thermal conductivity	watt per meter kelvin (W/m.K)	$m.kg.s^{-3}K^{-1}$
Energy density	joule per cubic meter (J/m³)	$m^{-1}.kg.s^{-2}$
Radian intensity	watt per steradian (W/sr)	$m^2.kg.s^{-3}/(m^2/m^2) = m^2.kg.s^{-3}$
Radiance	watt per square meter steradian (W/m².sr)	$(m^2/m^2).kg.s^{-3} = kg.s^{-3}$

These are only a handful of the phenomena of Nature described by a few of the base units of the SI. All the derived quantities listed above, except thermal conductivity, involve only the base quantities length (meter), mass (kilogram), and time (second); so these phenomena could, in principle, have been identified by the *savants* who created the metric system in 1795 who had the meter, kilogram, and second. They would not have had the kelvin as the base unit of temperature, as temperature was not included into the SI as a base unit until 1954, the 18th-century *savants* who created the metric system would have used the Centigrade scale of temperature. With only the meter, the kilogram and the second we can define energy, the driving force of Nature. Gottfried Leibnitz had pointed out that what he termed the *vis visa* of a body (kinetic energy) was proportional to the product of the body's mass and the square of its velocity; that is, $kg.m^2s^{-2}$ (see Table 9.3). Likewise, Isaac Newton had said that a force (F) that causes a body to change its speed is equal to the mass (kg) of the body multiplied by the acceleration (m/s²); that is, $F = m.a = kg.m.s^{-2}$, which is the definition of the newton in Table 9.2.

By using the laws of physics as a grammar, and the base units as expressions or words, we may construct a language that allows us to make predictions about phenomena that have not yet been identified, but which should be observable. Looking at the above tables, a scientist could, for example, ask questions such as: What happens if a force tries to twist a body instead of pushing (repelling) or pulling (attracting) it?; What happens if a force acts upon an area, not simply along a line?; or Is there a real, measurable phenomenon that arises when one couples the next highest power of length with time? In the first case, one defines torque, which is a force that tries to rotate a body (as any inexperienced motorcyclist, who has applied his rear-break too harshly while going around a corner too quickly will tell you). In the SI, torque is termed angular velocity (see Table

9.4) and is expressed in radian per second. But as the radian is dimensionless in the SI, it is written as per second (s^{-1}). This can be confusing and is the reason why many engineers prefer to express this important quantity in non-SI units. A force acting over an area would be newtons per square meter, and from the Tables we see that this quantity would be $m.kg.s^{-2}/m^2$ or $m^{-1}.kg.s^{-2}$, which is the definition of pressure. Pressure is nothing more than the force exerted by something (a gas or a fluid) upon a well-defined area.[21]

As for the reasonableness of phenomena that have not yet been observed, as was mentioned above, one first has to consider the magnitude of the units and a dimensional analysis to see if such a new phenomenon is observable. A relatively recent example would be the pressure exerted by light, that is, radiation pressure. Could it exist? Is it measureable? The answer was yes, and it was discovered that the radiation of the Sun exerts a pressure of less than a billionth of an atmosphere at the Earth's surface. But it was an examination of the existing language of science, which suggested to and allowed individuals to look for this new phenomenon. The complex interconnectedness of the base units that couple, to generate the phenomena of Nature is represented in Figure 9.2. This figure tells us, for example, that the present definition of the second is used in the present definition of kilogram, meter, candela, ampere, and kelvin, and that the definition of the unit of temperature, the kelvin is dependent upon mass, time, and length. This organic wholeness of the phenomena of Nature reminds us of the ideas underlying the *I Ching* and the Taoist view of Nature. [2]

In linguistics, grammar is the set of rules governing the composition of clauses, phrases, and strings of words in any given natural or vernacular language. Individuals who use or speak a language have a set of internalized rules for using that language, and these rules constitute that language's grammar. The vast majority of the information in the grammar is, at least in the case of one's native language, acquired not by conscious study or instruction, but by observing other speakers. Much of this work is done during early childhood; learning a language later in life usually involves a greater degree of explicit instruction. But for all the emotion expended by those who understand and use the rules of grammar, as well as those who have no idea about grammar; grammar is the cognitive information underlying language use. And this is the same in the language of science, as it is in English. Grammar allows us to turn the lists of verbs, nouns, adjectives, adverbs, etc. that come into our minds, at a particular moment into comprehensive, information-conveying prose. Grammar is not a distraction or an irritation; grammar is magical. Indeed, grammar and grimoire are derived from the same root. Grammar is also glamour, and the primary meaning of glamour is enchantment or spell. While grimoire is a manual for the casting of spells. Through grammar we may define, explore, understand, and perhaps control some aspects of the Universe.

---

[21] The British customary unit for pressure (still used in, for example, tyre pressures) is pounds per square inch, which gives clear indication of pressure as a force upon an area.

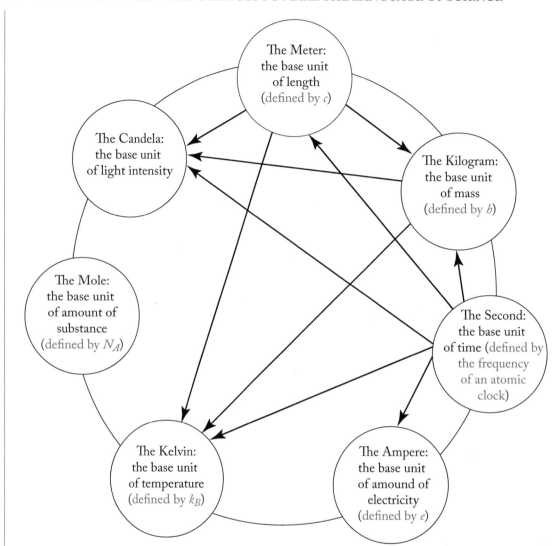

Figure 9.2: The interconnectedness of the seven base units of the SI, now that (since May 2019) the kilogram is redefined via Planck's constant, h, the unit of thermodynamic temperature defined by Boltzmann's constant, k_B, the mole defined by Avogadro's number or constant, N_A, and the ampere defined by the charge of the electron, e. As can be seen, the network of connections is complex, for example, the kelvin is connected to the physics underlying the SI and is defined as an energy, but is dependent on the definitions of length (L), mass (M), and time (T) as $\mathsf{M.L^2.T^{-2}}$. In addition, the ampere is defined as a flow of electrons in a time interval and so is no longer a force ($\mathsf{M.L.T^{-2}}$) dependent on length, mass, and time, but only on time. The kilogram is now defined by h, which is dependent on energy and time.

9.3   LOCATION: THE SURFACE OF MARS, SEPTEMBER 23, 1999

On the September 23, 1999, the Mars Climate Orbiter satellite was lost during a maneuver to place it in an orbit around Mars. Instead of entering a stable orbit from where it could monitor the Martian weather, it is believed that the satellite crashed onto the surface of the Red Planet. After the long crossing of interplanetary space, the satellite's controllers would have needed to slow down the satellite for it to safely enter the Martian atmosphere. It is believed that it was this braking process, which led to the satellite's loss. To slow down a body in motion requires the application of a force of the same order of magnitude as the velocity of the satellite, but applied in the opposite direction. Unfortunately for the satellite, and the scientists waiting to hear from it, something went seriously wrong.

As it turned out, the problem for the Mars Climate Orbiter was that the force necessary to slow the satellite down for it to safely enter a stable orbit around Mars, was calculated in one set of units, but when the command was sent to the satellite to ignite the braking thrusters, it was applied in a different set of units. The two sets of software, on Earth and on the satellite hurtling toward Mars, were trying to communicate in different scientific units, and instead of entering a stable orbit well above the surface of the planet, it attempted to enter an orbit much closer to the surface and crashed.

The principal cause of the disaster was traced to a thruster calibration table, in which British customary units instead of metric SI units had been used to define force. The navigation software expected the thruster impulse data to be expressed in newton seconds, but the satellite provided the values in pound-force seconds (a non-SI unit). This confusion in units caused the electric impulse to be interpreted as roughly one-fourth its actual value. Knowing the forward-moving force of the satellite, a calculation would have been undertaken to determine the force require to be applied in the reverse direction to reduce the forward force, but due to incompatible programming on earth and in the satellite, the satellite crashed.

The pound-force (lb_F) is a unit of force in the system of units, loosely term British customary units. There are many ways to define force in this system of units, which may be considered confusing to some, but actually tells one a great deal about the physics going on in a particular experimental situation. The pound-force is equal to the force exerted on a mass of one *avoirdupois* pound on the surface of Earth. Originally, this unit was used in low-precision measurements where small changes in the Earth's gravity (which varies from place to place on the surface of the Earth by up to half a percent could be neglected. The acceleration of the standard gravitational field (g) and the international *avoirdupois* pound (lb_m) define the pound-force as: $1lb_F = 1lb_m \cdot g = 1lb_m \cdot 32.174$ feet per second squared, which on converting to the SI is equal to 0.454 kilogram . 9.806 65 meter per second squared = 4.448 newton. (This factor of 4.448 was absent from the software that controlled the satellite, and so it crashed.)

Even after 200 years of the decimal metric system, the world of science and technology is still full of different systems of units. That is, different dialects of the same universal language of science. But provided one is aware of these different dialects, one may affect the necessary translation in a trivial line of computer code and everything will be fine. Assuming, blindly, that everyone is speaking exactly the same dialect by assuming that everyone is using the same system of units is risky.

9.4   A FINAL COMMENT ON THE VALUE OF A QUANTITY: SACRED GEOMETRIES

The Book of Revelations is the *Ur*-text of a great deal of the nonsense one finds on the Internet. One example of the many Hermetic, or opaque subjects arising from Revelations often discussed at interminable length, but rarely with any clarity on the Internet are sacred geometries, or sacred architecture. Interestingly, these arcane concepts also reveal something of the nature of the way scientists look at the world.

The concept of a sacred geometry, or a sacred architecture, comes from Saint John's vision of the Heavenly Jerusalem given in Revelations 21:17. The King James Bible tells us about the dimensions of the future New Jerusalem, "*And he measured the wall thereof—an hundred and forty and four cubits, according to the measure of man, that is, of the angel.*" The cubit was a unit of length measurement common to the Ancient Mediterranean world, equal to the length of a man's forearm (elbow to finger-tip). We immediately see from this obscure text from Revelations that the author is in fact referring directly to the 5th century BCE, pre-Socratic Greek philosopher Protagoras' well-known comment that "*man is the measure of all things,*" but we are also given the actual height of the walls of the heavenly city, 144 cubits. This dimension has over the last two millennia, inspired many individuals, particularly, architects.

The great gothic Cathedral of Amiens in northern France, where construction began in 1220, is built to a height of 144 *pieds Romans* (that is, 42.3 meters or 138.8 British feet). The near-by Cathedral of Beauvais, on the other hand, where construction began in 1225, is built to a height of 144 *pieds du Roi* (that is, 48.5 meters or 159 feet). These two mediaeval units, Roman feet (*pieds Romans*) and Royal feet (*pieds du Roi*), are different. The fact that Royal feet are longer than Roman feet means that the Cathedral of Beauvais is higher than the Cathedral of Amiens, which may well explain why it was an unstable building that partially collapsed in 1284, while the Cathedral of Amiens has never fallen down.

As far as the mediaeval architects of these two neighboring cathedrals were concerned, it was the numerical value of the height of the City of God that was important. One hundred and forty-four units was to be the height of the cathedrals, because that was the height of the City of God given by Saint John in Revelations 21:17; it appears not to have mattered much which units was actually adopted. (Remember: any value of a physical quantity, Q, may be expressed as the product of a unit [Q] and a purely numerical factor or number.) The gothic architects of northern France were only interested in the numerical factor of this physical quantity taken from the Book of Revelations; as far as they were concerned, the unit was irrelevant. This addiction to the fetish value of a number is numerology, pure and simple; it is Hermeticism. That is, attempting to find some meaning or, perhaps, a secret hidden in a particular number. What precisely is the meaning of 144? Perhaps it is related to other celebrated biblical numbers such as 666? What is it about 144; the height of the walls and the number of the just (144,000)? When you put the heights of the two cathedrals into British feet or meters, any mystical significance or magic disappears in the very different light of the English Industrial Revolution and the French Enlightenment.

Conversely from the architects of mediaeval France, the Ancient Sumerians had little concept of pure numbers, although they were well versed in the use of numbers to hide mystical significance and to work magic. Our earliest recorded list of objects comes from Ancient Sumeria, and when we look at these ancient lists (more than five millennia old) we see that the scribes used the same metrological symbol, or unit as many times as was required by the value of the numerical factor before the unit. Thus, instead of writing six oxen as the Sumerian form of the number six followed by, perhaps, a schematic of an ox, the scribes simply drew the pictorial schematic of the ox six times.

The Ancient Sumerians used only the unit part of the definition of a quantity, while the mediaeval French craftsmen only used the numerical part of the definition of a quantity. Neither is the correct approach. One can readily imagine the evolution of the Sumerian usage to a more sensible, modern approach arising because the lists were getting to be very long and tedious to compose, and those clay tablets were so very small. The magico-Christian architects, on the other hand, got caught up in looking for mystical significance in the numbers mentioned in religious-poetic texts of late antiquity, and as a consequence they lost their way in numerology, and like the Tower of Babel their cathedral collapsed.

9.5 FURTHER READING

1   Concerning the origins, since the late 19th century and the previous definitions of the base units of the SI, the best source for detailed information is the bilingual *SI Brochure* published by the *Bureau international des poids et mesures* (BIPM). This is a non-technical document intended for those already familiar with the science relating to the origin of the SI, and is essentially a list of rules concerning the use of SI units. The brochure is written by the Consultative Committee for Units of the BIPM; the most recent edition, the 8th, was published in May 2006. This substantial booklet is only available through the BIPM, but the text is freely available on the BIPM's website (www.bipm.org/en/publications/). Also included are extensive lists of references as to when certain words or quantities were adopted for use with the SI.

The BIPM website also contains freely available information about the evolution of the definitions of the base units of the SI. The definitions of several of the base units changed in May 2019—full details may be found on the BIPM website (in English and in the official language, French) at https://www.bipm.org/en/measurement-units/.

Also see https://en.wikipedia.org/wiki/2019_redefinition_of_the_SI_base_units.

2   There is also a more recent history of metrology and of the metric system in this volume: *Defining and Measuring Nature: The Make of all Things* (2014); Jeffrey H. Williams; San Rafael, CA, Morgan & Claypool.

CHAPTER 10

What Powers Society?

We already know enough to begin to cope with all the major problems that are now threatening human life, and much of the rest of life on earth. Our crisis is not a crisis of information; it is a crisis of decision.

George Wald (1906–1997)

Having looked briefly at how the worldview of scientists has evolved, let us now briefly begin to consider the mutual influence of science and society. In particular, we will do this by looking at bulk thermodynamic properties (see Table 10.1, which is based on the tables in Chapter 9). In physics, energy—derived from the Greek ἐνέργεια (activity or operation) a term that first appears in Aristotle—is an indirectly observable quantity. Energy powers, not only us and our society, but also the Universe and everything in the Universe; however, it cannot be measured as an absolute quantity. We may say that there is energy in a system, but we cannot quantify it precisely; as Saint Thomas Aquinas said in the 13th century, "*I see motion, so I infer energy.*" Indeed, there is so much energy in the Universe that we are only really able to quantify the influence of changes in quantities of energy, as they act upon matter. Table 10.1 lists four derived units from the International System of Units (see Chapter 9), for energy, work, force, power, and pressure. We can see from their representation in the language of science that they only differ by powers of length (m in meters) and time (s in seconds). The final column also demonstrates how interconnected are these four basic phenomena— as revealed in Figure 9.2.

Table 10.1: Named derived closely related quantities		
Quantity	Derived Unit (Symbol)	Representation in the Language of Science
Energy (and Work)	joule (J)	$kg\ m^2\ s^{-2}$
Force	newton (N)	$kg\ m\ s^{-2}$
Power	watt (W)	$kg\ m^2 s^{-3}$
Pressure	pascal (Pa)	$Nm^{-2} = kg\ m^{-1}s^{-2} = Jm^{-3}$

James Prescott Joule (1818–1889), Figure 10.1, was an English brewer and amateur scientist, born in Salford, Lancashire. Joule studied the nature of heat, and discovered its relationship to mechanical work. This led to the law of conservation of energy, which in turn led to the development of the first Law of Thermodynamics. The SI-derived unit of energy, the joule, is named after him.

James Watt (1736–1819), Figure 10.2, was a Scottish inventor, mechanical engineer, and chemist who improved on Thomas Newcomen's 1712 Newcomen steam engine with his Watt steam engine of 1776, which went on to become the driving force of the Industrial Revolution around the world.

Figure 10.1: A photograph of James Prescott Joule, the Salford brewer and amateur scientist who defined and quantified energy. Image from: https://en.wikipedia.org/wiki/James_Prescott_Joule#/media/File:Joule_James_sitting.jpg.

When we look up at the sublime spectacle that is the night sky, all we see is energy and matter, nothing else, and as Einstein pointed out, energy and matter are proportional, and can be equated with a constant equaling the speed of light (c) squared ($E = m.c^2$). All the things we see around us—form, texture, color, together with the sublimity and sense of the numinous that arise in our minds when we contemplate the Universe—all arise from the distribution of energy. But static quantities of energy and matter; stationary for all eternity can do nothing. It is only when energy and matter vary in quantities with distance that we can conceive of a Universe such as ours; a Universe capable of supporting life. If the Universe were not dynamic, in fact expanding, it would be a dark, dead gaseous nothing. In this, the Universe is like an economy. If we all kept our money in banks and spent nothing, there would be no global economy. Economic activity that generates jobs and opportunities, together with economic growth arises from the movement of money. That is, by money moving from one location to another. If it all stayed put in the bank accounts of a few billionaires, there would be no economic activity. Money only generates something useful to the wider society when it moves. The same with energy in a system. Everything we see around us, especially life, comes from the flow of energy.[22]

When energy is channeled into a particular direction, we have a force; a force is energy/distance, that is, $kg.m^2.s^{-2}/m = kg.m.s^{-2}$ as in Table 10.1. Again, the same can be said of money. When money flows into an economy, it can be said to be a powerful force. Consider a definition: *a force is any influence that causes an object to undergo a change in its motion* (for example its velocity or speed), *the direction of its motion, or its shape* (for example, compression). A force may thus cause a moving object to change its velocity (the force of gravity defining the trajectory of comets moving around the Sun; or an injection of capital to change the direction of an economy) or even to begin to move if it were stationary (the force of the rocket engine accelerating against the force of gravity, or motion in a game of billiards, or a start-up grant) or it may compress the shape of an object (collapse or bankruptcy). While mechanical stress can often remain embedded in solids, gradually deforming them, mechanical stress in a fluid gives rise to changes in the fluid's pressure if the volume is constrained.

Work, energy, power, and force are all closely related, indeed, they are interconnected in physics (see Table 10.1). But they are also closely related concepts in the wider society. Work is described as the product of a force multiplied by the distance over which it acts. Only the component of a force (which is actually a field of potential action extending in many directions) in the direction of the movement of its point of application does work. The term work was first used in 1826 by the French mathematician Gaspard-Gustave Coriolis (1792–1843), who gave his name to the force responsible for the swirling form of hurricanes, and of the vortex of water as it disappears down a plughole.

If a constant force of magnitude F acts on a point that moves a distance *l* in the direction of the force, then the work W done by this force is calculated from W=F.*l*. In the SI system of units, force is measured in the newton and it would act over a distance in meters, so the work done would equal newton meters, which (see Table 10.1) would be an energy, and so would be equal to joules. In the SI, work and energy have the same units.

Power is the rate at which energy is transferred, used, or transformed into another form of energy. For example, the rate at which an electric heater transforms electrical energy (electrical current per unit time) into heat and light by passing the current through a heating element of high resistance, and is measured in watts, in honor of James Watt (Figure 10.2). The greater the power output or wattage, the more power, or equivalently the more electrical energy is used per unit time. Thus, power is the time-averaged use of energy ($kg.m^2.s^{-2}/s = kg.m^2.s^{-3}$, as in Table 10.1).

---

[22] As the British Nobel laureate in biochemistry, Frederick Gowland Hopkins (1861–1947) commented, "*Life is a dynamic equilibrium in a polyphasic system.*"

Figure 10.2: James Watt painted by Carl Frederik von Breda. Image from: https://upload.wikimedia.org/wikipedia/commons/1/15/Watt_James_von_Breda.jpg.

Energy transfer can be used to do work, so power is also the rate at which this work is performed. The output power of an electric motor is the product of the torque the motor generates and the angular velocity of its output shaft. The power expended to move a vehicle is the product of the traction force of the wheels against the ground and the velocity of the vehicle. The SI unit of power is the watt, which is equal to one joule per second. Older, more picturesque units of power include ergs per second, horsepower, metric horsepower (in German, *Pferdestärke*), and foot-pounds per minute. One horsepower is equivalent to 33,000 foot-pounds per minute, or the power required to lift 550 pounds of weight by one foot in one second, and is equivalent to about 746 watts (and has nothing to do with carts and horses).

10.1   SOCIAL FORCES

By the late 19th century, classical physics was triumphant. We understood how the Universe functioned because we understood how energy was transfered, conserved, and partitioned in our laboratories, particularly, the Cavendish Laboratory in Cambridge, and we merely extrapolated to the larger scale of the Universe. In 1864, Maxwell published *A Dynamical Theory of the Electromagnetic Field*, where he first proposed that light was composed of waves moving in the same medium that gives rise to electric and magnetic forces. Maxwell's work in electromagnetism has been called the second great unification in physics, after the first great unification achieved by Isaac Newton. Maxwell wrote, "*The agreement of the results seems to show that light and* [electro] *magnetism are affections of the same substance, and that light is an electromagnetic disturbance propagated through the field according to electromagnetic laws.*" Maxwell was proved right, and his quantitative connection between light and electromagnetism is considered one of the great accomplishments of the 19th century in any field of endeavor (see Section 12.2).

By considering the propagation of electromagnetic radiation as a field emanating from some active source, Maxwell was able to advance his work on the nature of light. And by estimating what the speed of light should be, and observing that his prediction agreed with the best available experimental values, he was able to say that his initial assumption had been correct. In this way, he laid the foundations of the modern scientific methodology of solving complex interrelated problems. Even though Maxwell reconciled electricity, magnetism, and light, he did not live long enough to finalize the details of the character of the electromagnetic field. At that time, Maxwell and many others believed that the propagation of light required a medium which could support the waves, and through which the waves could move or propagate (as was the case for sound waves in air—sound waves cannot cross a vacuum). This proposed medium was called the *luminiferous aether*. Over time, however, the existence of such a medium permeating the Universe, and yet apparently undetectable by any mechanical means, proved more and more difficult to reconcile with experiment. Moreover, it seemed to require an absolute frame of reference in which the equations were valid, with the extraordinary result that the equations governing the phenomena of electromagnetism and optics would be different for a moving observer and for an observer at rest. These difficulties inspired Albert Einstein to formulate the theory of special relativity, and in the process Einstein demonstrated that one could dispense with the requirement for a sustaining *luminiferous aether*.

The scientists of the late 19th century spoke about energy as the means of powering the Universe, and of forces operating everywhere throughout the Universe; they appeared to possess near-divine competence to explain and predict what was going on here on Earth as well as in the distant reaches of the Cosmos. The period from the mid-19th century to World War I was the great period of scientific triumphalism, and it is not surprising that this attitude of authority was copied by many social scientists and those involved in ordering and maintaining society. After all, the application of a quantitative way of looking at life was a wide-ranging consequence of the French Revolution; we were all now subject (whether we knew it or not) to a new tyranny, the tyranny of numbers and scientific concepts. Society was to be run efficiently, like a late 19th century laboratory. This quantitative view of society and of the evolution of society is best represented in the work of the German philosopher and economist, Karl Marx, especially his *Das Kapital*, published posthumously in 1885 and 1894.

For those interested in the history of science, the concept of power is now inseparable from politics and economics. This confusion of terms began in the late 18th century when Matthew

Boulton, the financier who supported James Watt's work to develop the steam engine that powered the Industrial Revolution wrote to Empress Catherine (the Great) of Russia, in an attempt to sell the new steam engines he was developing with Watt. He wrote to Her Imperial Majesty, "*I am selling what the whole world wants: power.*" He had given the game away. That was what it was all about. That a discussion of power in Nature is really inseparable from the idea of power in society. We could say that whereas the fundamental principle of physics is energy, power is the fundamental principle of the social sciences, and of politics and economics. Perhaps because power can be readily quantified, if only by demonstrating that one politician can command more votes than another, or that one bank, organization, or media Moghul is richer than another bank, organization or media Moghul, that they are deemed to be more powerful.

Corporate power may have a different sense from physical power, but the units are exactly the same, and we have a confusion of terms. Such homonyms, are known in other areas. The ideas presented here are an attempt to point out that our society is really a microcosm of Nature, and concepts of what makes the Universe work the way it does, can be applied (perhaps without too much difficulty) to our society. Energy is the currency of Nature, and power and force drive nature; in society, money may be currency, but it is also the medium, the power and force that brings about change in society.

When we come to forces, we are in an even more difficult position about possible confusion between politics, the social sciences, and the physical sciences. How many times have we heard a partisan journalist say of a politician that "he/she was a force of Nature." A force is something that compels a molecule, or a planet to do something; that is, it is energy directed along a well-defined path, so perhaps certain politicians are able to act as forces. They order people to do this or that, they change society and use considerable resources in their endeavors. However, the big difference is that in Nature forces act in the most efficient manner, there is little wastage of energy; politicians, on the other hand, are less efficient and waste a great deal of the scarce currency of society.

The age of scientific triumphalism came crashing down with the World War I, the advent of quantum mechanics and relativity in physics, and with modernism in the arts (cubism in the visual arts did more to advance the ideas of Einstein and Heisenberg than any number of textbooks written for a general readership). Sadly, the quantitative triumphalism of the social scientists and of politicians took a bit longer to dissipate, but another world war and several economic crashes; particularly, that of 2008, have revealed the final bankruptcy of all the standard models of economic growth; yet politicians still believe that we can all continue to have unlimited economic growth in a closed system of finite resources, that is, it is possible for politicians to repeal the Second Law of Thermodynamics. The force of politics has gone, and all we have left are politicians seen as a disruptive or, at best, an irrelevant force in social progress.

10.2   INTERNATIONAL REGULATION OF TERMS AND NAMES: DIALECTS ARE INEVITABLE

I have spoken about the scientists' desire to create a simple language to describe the world that would be universally understood; to permit a return to a Golden Age. But are such dreams practicable in the wider society? The followers of such projects always try, with greater or lesser cohesive power, to realize an international forum. But which authority has the competence to adjudicate between these contending parties? Is it the richest or the most powerful nation on earth that decides for the rest of humanity? The beginning of the last century was the most optimistic epoch for the creation of Utopic ideas of international committees deciding on matters affecting and effecting humanity. This was an epoch when it still seemed realistic to believe that an international body would be capable of coming to a fair and ecumenical conclusion, and imposing it on every nation by reason. But two world wars and numerous economic depressions put an end to all that Utopian nonsense.

Anyway, if a committee did make a useful contribution; for example, inventing a good, new candidate for a universal language; as soon as it was made public, the language would spread through various countries. There would be clubs to propagate this new language, and these clubs would begin petitioning national governments to access national education systems. However, what invariably happens is that the original inventor discovers that his/her language has been subjected to, supposedly "heretical" modification(s), which might further simplify, restructure, and rearrange it—making it more useful as an international language. But the original inventor for, whatever reason will likely not be happy about this. The product of all their labor will have been modified by others. Their creation is not the final version. That honor will go to someone else. Such will inevitably be the fate of artificial languages: the "word'" remains pure only if it does not spread; if it spreads, it becomes the property of the community of its proselytes, and (since the best is the enemy of the good) the result is "Babelization." After a few short years of rapid inflationary growth, the movement collapses, and continues only in an ever-shrinking state.

One may make the observation that a universal language is impossible for a simple reason: Even if everybody on earth agreed to speak the same language from today, they would soon discover that, under the influence of their own use the single languag,e had begun to change, to modify itself in a multitude of different ways in each country, until it produced in each a different dialect, which gradually grew away from all the others. It is for this reason that the Portuguese spoken today in Brazil differs from the Portuguese spoken in Portugal and, more famously, the ever-widening separation of English spoken in the UK and in the U.S.

10.3   SCIENCE AS A NEW TOWER OF BABEL

By this point, I am sure that the reader will have appreciated that creating a universal language of science, or even a new system of weights and measures is no easy matter. Even ignoring political conflicts, it is rarely possible to achieve consensus between relatively small groups of scientists as to which units they should be using. And as for devising a scientifically coherent system of units that may be adopted and used by the wider society there is no simple answer.

Creating a system of weights and measures, or a universal language capable of quantitative extension, is difficult due to conflicting requirements.

- To facilitate everyday use, the units or nomenclature should be of a size or facility that is appropriate for use in specialist areas of science and technology, but they must also be appropriate for everyday use by the wider community.

- To facilitate international use, the units or nomenclature should be defined in a manner that is both precise and capable of being reproduced anywhere in the world, and not be subject to reference to a prototype or artifact kept by a particular nation.

- The units or nomenclature should be coherent; that is, all the subsidiary or derived units, which are needed to fully describe Nature, can be expressed as combinations of the basic units without the introduction of any numerical constants.

Fortunately, compromise and even pragmatism are not entirely unknown in science, and some progress has been possible. The centimeter-gram-second (CGS) system of units was widely used until the electricity industry decided that the electromagnetic unit, or emu derived from Ampère's Force Law, gave quantities which were too small for practical use by electrical engineers. They rescaled the electrical units to make them more "user friendly" for the electricity generation and supply companies, who needed units which referred directly to the large electric currents and the huge voltages found in industry, rather than the much smaller values used by research scientists. This resizing of units made the electrical engineers happy, but coherence with mechanical units was lost because of the numerical factors which were now needed to connect the units for large currents and voltages to related quantities. In 1948 it was decided that the centimeter, the gram and the erg (the CGS unit of energy) should be replaced by the larger meter, kilogram, and joule (one centimeter equals 10^{-2} meters, one gram equals 10^{-3} kilogram, and one erg equals 10^{-7} joules). This new system of units was called the meter-kilogram-second-ampere (MKSA) system of units, and it restored coherence to the whole system of units and quantities; that is, the numerical conversion factors introduced by electrical engineers disappeared. The MKSA system of units is also known as the SI system of units (see Chapter 9).

So, why use these different systems of electrical units? Why was it deemed necessary to appease the electricity industry? Well, it is not simply a reluctance to change. It is all about what you hold to be important. Pragmatists favor the SI's utilitarian approach to calculation, which actually keeps a lot of the underlying complex physics out of sight (which is useful for teachers when trying to instruct bored students, but is inappropriate for researchers); on the other hand, philosophically minded physicists want only the base units to better reflect the underlying science.

This problem of fundamentally different systems of units and nomenclature being used concurrently by different communities of both scientists and non-scientists does, of course, extend far beyond the world of electrical engineering. A similar "confusion of tongues" applies in something as technically straightforward as pressure measurement. (Pressure is defined as the force per unit area applied in a direction perpendicular to the surface of a vessel. It can be thought of as arising from the molecules of gas striking the inner surface of the vessel containing that gas.)

The SI unit for pressure is the pascal (Pa), named in honor of the 17th-century French mathematician and Catholic philosopher Blaise Pascal (1623–1662), which in the SI is equal to one newton per square meter (that is, N/m^2 or $kg.m^{-1}.s^{-2}$; pressure is a force). The name "pascal" (symbol Pa; see Table 10.1) for the unit was adopted in 1971; before that date, pressure in the SI was expressed simply as so many N/m^2. The problems associated with pressure measurements begin with the SI unit of area; one square meter is a very large area, and so the values of even the modest pressures encountered in everyday life are very large numbers. For example, the pressure in your car tyre would be about 340,000 Pa; and successful systems of units usually express everyday quantities in small numbers, so as to facilitate familiarity with the size, and in recording and quoting the values (particularly, with non-physicists in places such as garages and tire shops).

On the other hand, non-SI units of pressure are legion. There are pounds per square inch (psi), or more precisely (given the distinction between mass and weight) pounds-force per square inch, and bars (that is, atmospheres) which are commonly used in the English-speaking world. The CGS unit of pressure is the barye (ba), equal to 1 dyn/cm^2 or 0.1 Pa (a dyne, abbreviated as dyn, being the unit of force in the CGS system of units, one dyne is equal to 10^{-5} newton). Then there is the universal measure of pressure used by the medical profession; your blood pressure of, for example, "130 over 82" is actually two measurements of pressure with each result given in millimeters of Mercury (mmHg). Here the pressure is defined as a force which would support a column of Mercury of uniform cross-section to that particular height.[23] The standard atmosphere (atm) is a well-known and well-used constant. It is approximately equal to the air pressure at sea level and is equal to 101,325 Pa or 101.325 kPa or 1013.25 hPa, in the SI, or 14.696 psi or 760 mm of Mercury; so 1 mm of Mercury is equal to 133.3 Pa.

---

[23] When millimeters of Mercury, or even inches of water, are quoted today as pressures, these units are not based on an actual physical column of Mercury or water; rather, they are measured by small electronic sensors or transducers whose readings could be calibrated or expressed in any number of units (SI or non-SI) by an inbuilt computer chip, calibrated to behave as if it were a column of Mercury or of water.

Another point to bear in mind about this profusion of units of something as straightforward as pressure is the difference between relative pressures (relative to atmospheric pressure) and absolute pressures (relative to a vacuum); for example, the pressure in your car tires is a relative pressure, or an overpressure and is often written as psig (pounds per square inch of gauge), which is one atmosphere or 14.696 psi above a vacuum. An absolute tire pressure would be written in psia (pounds per square inch absolute), and would be lower than a measurement of the same pressure given in psig by the amount 14.696 psi.

These are all expressions of the same piece of information, but expressed in the various dialects of the single language of science. This plethora of units for something as basic as pressure measurement is mirrored in measurements of many other common phenomena. This variety is not something that exists to confuse students and non-scientists. Such varieties of units exist for sound technical reasons: convenience in specialist branches of science, or convenience or facility of use in certain ranges of pressure, or because one profession refuses to change to another system of units, or because there is such an investment in technology that any change would be too expensive. The medical profession will, for example, not move away from using mmHg for blood pressure measurement,[24] which is convenient for them and a sufficiently precise measurement for their patients, but this is not the case for the vast majority of physicists who gave up using mmHg as a unit for pressure early in the last century.

However, the question we have to ask ourselves is whether there is anything to be gained by attempting to force a large body of professionals to give up a system of units with which they have become familiar over many generations. There is certainly the possibility of serious adverse consequences arising from such a move. It would be far better to encourage the ability to use and convert between many of these systems of units—to celebrate the diversity of the dialects of the single language of science. A scientist or a technician who can convert between these units will be someone who will truly understand the science underlying the phenomenon, and will be less likely to make foolish errors.

Any common language for science would inevitably and rapidly grow distant from the language of literature, but we know that the language of science and the language of letters influence each other. But, in addition, an international language of purely scientific communication would soon become an instrument of secrecy, from which the humble speakers of their own native dialects, or regional languages would be excluded. And as to possible literary uses, if the authors were obliged to write in a common tongue, they would be exposed to international rivalries, fearing invidious comparisons with the works of foreign writers. Thus, it seems that circumspection was a disadvantage for science and an advantage for literature, as it was for the astute and cultivated traveler, more learned than his native and naïve interlocutors. In the background to the formation of a universal

---

[24] In France, doctors define the blood pressure of their patients in terms of centimeters of Mercury, rather than millimeters of Mercury. That is only a factor ten to consider.

language there is an 18th-century prejudice, which is still with us; that people simply do not wish to learn other languages, be they universal or merely foreign. There exists a sort of cultural deafness when faced with polyglottism, a deafness that continued on throughout the 19th century to leave visible traces in our own time.

10.4   FURTHER READING

Defining and Measuring Nature: The Make of all Things (2014); Jeffrey H. Williams; San Rafael, CA, Morgan & Claypool.

Order from Force: A Natural History of the Vacuum (2015); Jeffrey H. Williams; San Rafael, CA, Morgan & Claypool.

CHAPTER 11

The Ghost of the Divine Language: The Theory of Everything

Reality is merely an illusion, albeit a very persistent one.

Albert Einstein (1879–1955)

The Theory of Everything (TOE) is the final theory, the ultimate theory, or master theory. It is a hypothetical single, all-encompassing, coherent theoretical framework of physics that will fully explain and link together all aspects of the Universe. Science fiction writers has long speculated about such an over-arching model of how the natural world functions, but let us consider what physicists mean by this fantastic, this sublime idea.

Finding a TOE is one of the major unsolved problems in physics. Over the past two centuries, two theoretical frameworks have been developed that, as a whole, most closely resemble a TOE. These two theories, upon which all modern physics rests are general relativity and quantum field theory. General relativity is a theoretical framework that only focuses on one of the four fundamental forces of Nature, gravity, for understanding the Universe on a large scale, and for objects of high mass: planets (the mass of the earth is about 6×10^{24} kg), stars, galaxies (the mass of a galaxy is estimated to be about 10^{42} kg), clusters of galaxies, etc. On the other hand, quantum field theory is a theoretical framework that focuses on the other three of the four fundamental forces of Nature (one of which is displayed in Figure 11.1), excluding gravity, and it holds for understanding the Universe at the small scale, and for objects of low mass: sub-atomic particles (the mass of a proton is $1.672\ 621\ 923\ 69(51) \times 10^{-27}$ kg), atoms, molecules, etc. Quantum field theory successfully implemented the Standard Model and unified the interactions (so-called Grand Unified Theory) between the three non-gravitational forces: strong nuclear force, weak nuclear force, and electromagnetic force.

By merely suggesting that there is, somewhere… at some energy, a TOE, and that when we discover it all experimental science will become redundant (that is the implication), we are again in the world of the mediaeval and pre-mediaeval *savants* who searched for the Perfect Language (see Chapter 2). Both the poet Dante and his exact contemporaries, the Kabbalists of Spain searched for the language used by God to bring the Universe into existence from nothing (*Ex nihilo*). They believed that this language was merely lost, or perhaps, given that it was a language of power, had been hidden from man. But once man had re-discovered this language, all the secrets of Nature would be revealed to him. He could re-order the world, and by implication humanity, bring about a new Golden Age of harmony, peace, and prosperity.

Figure 11.1: Lightning or electromagnetism in action. Lightning is probably the most spectacular, frightening, and immediate demonstration of one of the four forces of Nature (electromagnetism).

A great many of the most eminent physicists have confirmed with precision experiments virtually every prediction made by the two theories of general relativity and quantum field theory—when used in their appropriate domains of applicability. In accordance with their findings, scientists have also learned that general relativity and quantum field theory, as they are currently formulated are mutually incompatible; they cannot both be right. Since the domains of applicability of general relativity and quantum field theory are so different, most situations require that only one of the two theories be used. As it turns out, this incompatibility between general relativity and quantum field theory is apparently only an issue in regions of extremely small scale and high mass, such as those that exist within a Black Hole, or in the early stages of the Universe. To resolve this conflict, a theoretical framework revealing a deeper underlying reality, unifying gravity with the other three fundamental interactions, must be discovered to integrate, harmoniously the physics of the very large and of the very small into a seamless whole: a single theory that, in principle, is capable of describing all phenomena, at all length and mass scales.

Today, it is string theory that has evolved into a candidate for this ultimate theory of the Universe, but not without limitations and controversy. String theory posits that at the beginning of the Universe (up to 10^{-43} second after the Big Bang, when the Universe was very small, and so at a very high temperature), the four fundamental forces were a single fundamental force. According to string theory, every particle in the Universe, at its most microscopic level (the Planck length scale), consists of varying combinations of vibrating strings with distinct patterns of vibration. String theory further claims that it is through these specific oscillations that a particle of a unique mass and charge is defined. Thus the electron (of mass, $m_e = 9.109\ 383\ 56(11) \times 10^{-31}$ kg and of charge, $1\ e = 1.602\ 176\ 620\ 8(98) \times 10^{-19}$ C $= 4.803\ 204\ 51(10) \times 10^{-10}$ esu) is a string vibrating one way, while the up-quark, of charge $= +(\frac{2}{3})\ e$ and mass $= (\frac{1}{3})m_e$, is a string vibrating in a different manner.

11.1   SOME BACKGROUND

In Ancient Greece, philosophers such as Democritus (c.460–c.370 BCE) speculated that the apparent diversity of observed phenomena was due to a single type of interaction, namely the ability of the most fundamental particles, termed atoms, to move and collide with each other in the void that existed between the indivisible, eternal atoms. Archimedes (c.287–c.212 BCE) was possibly the first natural philosopher known to have described Nature with axioms (or principles), and then to have deduce new results from observations of these principles.

In the late 17th century, Isaac Newton's description of the force of gravity, which he knew operated over vast, astronomical distances implied that not all forces in Nature result from things coming into contact, or colliding. In his *Mathematical Principles of Natural Philosophy* of 1687, Newton gave us an example of the unification of physical principles; in this case, unifying the mechanics of Galileo Galilei on terrestrial gravity, the laws of planetary motion of Nicolaus Copernicus (1473–1543) and the phenomenon of tides by explaining these apparent actions at a distance under a single law: the Law of Universal Gravitation.[25] The mid 19th century saw the unification of electrical and magnetic phenomenon to create electromagnetism. In his experiments of 1849–50, Michael Faraday was the first to search for a unification of gravity with electricity and magnetism, but he was unsuccessful. In 1900, the German mathematician David Hilbert (1862–1943) published a list of mathematical problems that became famous, and stimulated a great deal of research in the early years of the last century. In Hilbert's sixth problem, he challenged researchers to find an axiomatic basis to all of physics. He asked the physics community to come up with a Theory of Everything.

In the late 1920s, the recently invented quantum mechanics showed that the bonds between atoms, which are the basis of all chemistry and physiology were examples of electromagnetic forces (see Figure 11.1, where the lightning results from the energy released by atoms and molecules excited and ionized by huge electric and magnetic fields). This discovery led one of the inventors of quantum mechanics, Paul Dirac, to boast in the Preface of the first edition of his textbook on

---

25 Newton's Law of Universal Gravitation states that every particle attracts every other particle with a force that is directly proportional to the product of their masses and inversely proportional to the square of the distance between their centers. This is a general physical law derived from empirical observations by what Isaac Newton called inductive reasoning. The equation for this law is: $F = G(m_1.m_2/r^2)$, where F is the gravitational force acting between two objects, m_1 and m_2 are the masses of those objects, r is the distance between the centers of those masses, and G is the gravitational constant (6.674×10^{-11} m^{3}kg^{-1}s^{-2}). We see immediately that this constant is known with considerably less precision that are the charge and mass of the electron given above.

quantum mechanics (*The Principles of Quantum Mechanics*; Cambridge University Press, 1930) that *"the underlying physical laws necessary for the mathematical theory of a large part of physics and the whole of chemistry are thus completely known."* Although Dirac predicted the end of experimental science, things turned out rather differently. During the great advances made in particle physics in the last century, that is, the elucidation of the menagerie of sub-atomic particles we know today, the search for a unifying theory was interrupted by the discovery of the strong and weak nuclear forces, which both differ from gravity and from electromagnetism.

Gravity and electromagnetism could always coexist as entries in a list of classical forces, but for many years it seemed that gravity could not even be incorporated into the quantum framework, let alone be unified with the other fundamental forces; the strong and the weak nuclear forces are purely quantum phenomena. For this reason, work on unification in the last century focused on understanding the three quantum forces: electromagnetism and the weak and strong nuclear forces. The first two were combined in 1967–68 by Sheldon Glashow (U.S., born 1932), Steven Weinberg (U.S., born 1933), and Abdus Salam (Pakistan, 1926–1996) into the electroweak force; see Figure 11.2. Electroweak unification is a broken symmetry; the electromagnetic and weak forces appear distinct at low-energies because the particles carrying the weak force, the W and Z bosons, have non-zero masses of 80.4 GeV/c^2 and 91.2 GeV/c2, respectively, whereas the photon, which carries the electromagnetic force, is without mass. At higher energies, the W and Z bosons can be created, and the unified nature of the force becomes apparent.

A TOE would unify all the fundamental interactions of Nature: gravitation, strong nuclear interaction, weak nuclear interaction, and electromagnetism. Because the weak interaction can transform elementary particles from one kind into another, the TOE should also yield a deeper understanding of the various kinds of sub-atomic particles. The usual assumed path of these various theories is given in Figure 11.2, where each unification step (indicated in Figure 11.2 with a short vertical arrow (↑) from two horizontal pre-existing theories) leads to a higher level of sophistication and complexity.

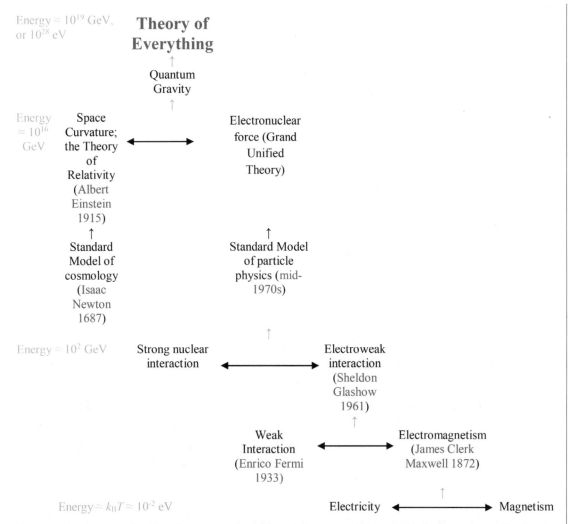

Figure 11.2: A generalized and theoretical route to the creation/discovery (whether science develops a theory to explain Nature, or discovers a theory that explains Nature is a moot point, but one beyond the scope of the present volume) of the Theory of Everything, or the Perfect Language needed to characterize and explain every detail and phenomenon of Nature. Originally, electricity and magnetism were considered to be two separate forces. This view was overthrown by James Clerk Maxwell in 1873. Albert Einstein published his general theory of relativity in 1915. In 1961, Sheldon Glashow combined the electromagnetic and weak interactions. In 1967, Steven Weinberg and Abdus Salam incorporated the Higgs mechanism into Glashow's electroweak interaction, giving it its modern form. The Standard Model was developed in stages throughout the latter-half of the 20th century, through the work of many scientists with the current formulation being finalized in the mid-1970s with experimental confirmation of the existence of quarks. A Grand Unified Theory (GUT) is a model in particle physics in which, at high energy the three gauge interactions of the Standard Model that define the electromagnetic, weak,

and strong interactions, or forces, are merged into a single force. Although this unified force has not been directly observed, the many GUT models theorize, or foretell its existence. If unification of these three interactions is possible, it raises the possibility that there was a grand unification epoch in the very early universe in which these three fundamental interactions were not yet distinct. The novel particles predicted by GUT models are expected to have extremely high masses of around the GUT energy scale of 10^{16} GeV; that is, just not far below the Planck scale of 10^{19} GeV, and so are well beyond the reach of any foreseen particle collider experiments. The total energy range covered by the physics in this figure is a staggering, 30 orders of magnitude (10^{-2} eV to 10^{28} eV).

The essential point to make about Figure 11.2, other than that it is a road map to an unknown destination (always the most exciting sort of journey); a destination that may not even exist, is that it represents a sequence of events at vastly different ranges of energy. From energies, of order, $k_B T$ for the unification of electricity and magnetism (here T is the temperature and k_B the Boltzmann constant that relates temperature to energy; electromagnetic phenomena occur at ambient conditions) to vast energies way beyond the capabilities of our present high-energy particle colliders. The electroweak unification occurs at around 10^2 GeV, Grand Unification is predicted to occur at 10^{16} GeV, and unification of the Grand Unified Theory force with gravity is expected at the Planck energy, roughly 10^{19} GeV (that is, about 1028 eV).[26]

Several Grand Unified Theories (GUTs) have been proposed to unify electromagnetism and the weak and strong nuclear forces. Grand unification would imply the existence of an electronuclear force; it is expected to set in at energies, of order, 10^{16} GeV, far greater than could be reached by any present particle accelerator. The Large Hadron Collider is the world's largest and most powerful particle collider, and the largest machine in the world. It was built by the European Organization for Nuclear Research between 1998 and 2008 in collaboration with over 10,000 scientists, and hundreds of universities and laboratories from more than 100 countries. It lies in a tunnel 27 km in circumference, 175 m beneath the France–Switzerland border near Geneva. First collisions were achieved in 2010 at an energy of 3.5 teraelectronvolts (TeV; 1 T eV = 10^{12} eV or 10^3 GeV) per beam, about four times the previous world record. After upgrades it reached 6.5 TeV per beam (13 TeV total collision energy, the present world record). At the end of 2018, it entered a two-year shutdown period for further upgrades; https://home.cern/.

---

[26] An electron volt (1 eV) is the amount of kinetic energy gained or lost by a single electron accelerating from rest through an electric potential difference of one volt in vacuum. Hence, it has a value of one volt, 1 Joule/Coulomb, multiplied by the electron's elementary charge e = 1.602 176 620 8(98) × 10^{-19} Coulomb. Therefore, one electron volt is equal to 1.602 176 620 8(98) × 10^{-19} Joule. By mass–energy equivalence, the electron volt is also a unit of mass (from Einstein's celebrated equation). It is common in particle physics, where units of mass and energy are often interchanged, to express mass in units of eV/c^2, where c is the speed of light in vacuum. The mass equivalent of 1 eV/c^2 is 1.782 × 10^{-36} kg. 1 eV corresponds to a temperature of about 11,604 K or 11,331°C. This system of units is useful for theoretical physicists, and the community of particle physicists but is wholly outside of the SI.

The final step in Figure 11.2 requires resolving the separation between quantum mechanics and gravitation, often equated with general relativity. Numerous researchers have concentrated their efforts on this specific step; nevertheless, no accepted theory of quantum gravity, and thus no accepted TOE have yet been formulated. In addition to explaining the forces mentioned in Figure 11.2, a TOE should also explain the status of, at least two candidate forces suggested by modern cosmology: an inflationary force for the Universe, and dark energy or dark matter.

11.2   STRING THEORY

We believe there are four fundamental forces that govern the Universe: gravity, electromagnetism, the weak force, responsible for beta-decay, and the strong force which binds quarks into protons and neutrons. The physics community believe they understand all of these forces except for gravity. The word "understand" is used loosely, in that we may define the Lagrangian,[27] which describes how these forces act upon matter and, at least in principle we know how to use these Lagrangians to make well-defined predictions with which to test theories. But with gravity, our understanding is incomplete. Certainly, we understand gravity classically; that is, in the non-quantum limit of $(h/2\pi)$ = 0, where h is Planck's constant. And provided we do not ask questions about how gravity behaves at very short distances (the Planck scale, of order, 10^{-35} m), we may calculate the effects of gravity.

It is sometimes said that physicists do not know how to combine quantum mechanics and gravity. In fact, physicists do understand how to include quantum mechanical effects into gravity, as long as we do not ask questions about what is going on at distances, less than the Planck length. For the other three fundamental forces we know how to include quantum effects, at all length scales. So, while we have a quantum mechanical understanding of gravity, we don't have a complete theory of quantum gravity. And that is the problem, as the most interesting questions we wish to ask about gravity is what happen at very small length scales; for example, questions such as "What was the Big Bang" and "What happens at the singularity of a Black Hole?" So what is it that goes wrong with gravity at scales shorter than the Planck-length? The mathematical answer is that the force of gravity is not renormalizable; that is, at very short length scales, the gravitational energy become very large (we are trying to divide something by zero), and we can only avoid this mathematical divergence (that is, a number tending toward ∞) by constructing unnatural models.

---

[27] Lagrangian mechanics is a reformulation of Newtonian mechanics, introduced by the Italian-French mathematician and astronomer Joseph-Louis Lagrange (1736–1813) in 1788. In Lagrangian mechanics, the trajectory of a system of particles is derived by solving the Lagrange equations in one of two forms; one separates terms involving kinetic and potential energy. In each case, a mathematical function called the Lagrangian is a function of the generalized coordinates, their time derivatives, and time; containing the information about the dynamics of the system. No new physics is necessarily introduced in applying Lagrangian mechanics, compared to Newtonian mechanics. It is, however, more sophisticated mathematically.

The force of gravity is a property of the scale over which it is being investigated. Consider an isolated electron in classical electromagnetism. The total energy (E_T) of this electron (of rest mass, m) is given by the sum of a kinetic part and a potential part:

$$E_T \approx m + \int d^3x \, |E|^2 \approx m + 4\pi \int r^2 dr \, (e^2/r^4),$$

where e is the charge of the electron and r is its radius. This integral defines the potential energy of the electron, and it diverges because you are dividing by a number going to zero at $r = 0$ (a point singularity). We may avoid this divergence by cutting the function off at some scale, Λ; so the total energy of the electron is now given by

$$E_T \sim m + C \, (e^2/\Lambda).$$

Clearly, the second term still dominates in the limit which interest us; that is, small r. Naively, we speak of the rest mass of the electron, m, but we cannot measure m; we measure E_T. That is, the inertial mass should include the electromagnetic self energy. Consequently, the physical mass m is given by the sum of the bare mass m and the mass derived from the electron's field energy (via $E = mc^2$). This means that the bare mass is infinite in the limit of interest. Note that we must make a measurement to fix the bare mass. We cannot predict the electron mass. It also means that the bare mass must cancel the field energy. That is, we have two huge numbers which cancel each other extremely precisely. To understand this better, note that it is natural to assume, using dimensional analysis (see Chapter 9) that the cut-off should be the Planck length. Which in turn means that the self-field energy is, of order, the Planck mass. So the bare mass must have a value which cancels the field energy. This cancellation is sometimes referred to as a hierarchy problem. This process of absorbing divergences in masses, or couplings (an analogous argument can be made for the charge e) is called renormalization.

String theory, however, replaces point particles with minute strings, which can be either open or closed (depending on the particular type of particle that is being replaced by the string), whose length, or string length (denoted l_s), is approximately 10^{-35} m. Moving from a point particle to a string avoids the problems of renormalisation. In string theory, one thus replaces Feynman diagrams by surfaces, and word-lines, or trajectories become world-sheets (see Figure 11.3). One increases the dimensionality of the problem.

All such theories use supersymmetry, which is a symmetry that relates elementary particles of one type of spin to another particle that differs by a half-unit of spin. These two partners are called superpartners. Thus, for every boson there exists its superpartner fermion and vice versa. For string theories to be physically consistent they require ten dimensions for space-time. However, our everyday world is only four-dimensional (three spatial dimensions and time) and so one is forced to assume that the extra six dimensions are extremely small, but must still be taken into consideration. To generalize about going from moving point particles (moving along a trajectory or world-line) to strings we have a world-sheet instead of a world-line, which has the form of a curved sheet or a curved cylinder depending on whether the string is open or closed (see Figure 11.3).

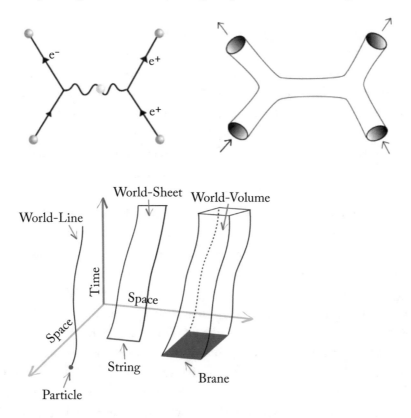

Figure 11.3: (upper) In string theory, Feynman diagrams (the one displayed here describes the scattering, or interaction of an electron, e⁻ and a positron, e⁺) involving point entities are replaced by surfaces. (lower) World-line, world-sheet, and world-volume (denoting the trajectories of particles, strings or branes), as they are derived for particles, strings, and branes.

What happens with gravity with this type of renormalization procedure? How does string theory solve the renormalization problem in gravity? Because the electron (now a string, not a point; one-dimensional rather than zero-dimensional) has finite extent l_p, the divergent integral is cut-off at $r = l_p$. We now have no need to introduce new parameters to absorb divergences, as they do not arise.

String theory has only one unknown parameter, the string length, of order, l_p, but this can be fixed by the only measurement string theory requires before it can be used to make predictions. However, the hierarchy problem remains. String theory predicts that the electron mass is huge, of order, the inverse length of the string, but we still need an additional something to give a reasonable value for the mass. It turns out that string theory can do more than just cut off the integral, it can also add an additional integral which cancels off a large part of the first diverging integral, leaving a more realistic result for the electron mass. This cancellation is a consequence of supersymmetry which, as it turns out is necessary in some form for string theory to be mathematically consistent. So by working with objects of finite extent as opposed to point particles, we accomplish two things. All the integrals are finite and in, principle, if string theory were completely understood we would only need one measurement to make predictions for gravitational interactions at arbitrary length-scales. In addition, we also gain predictive power (at least, in principle). Indeed in the Standard Model of particle physics, which correctly describes all interactions to energies, of order, 200 GeV, there are 23 free parameters which need to be fixed by experiment; for example, the electron mass. String theory, however, has only one such parameter in its Lagrangian, the string length.

However, one must never forget that physics is a predictive science; it is not an end in itself. The less descriptive and the more predictive a theory becomes the better. In that sense, string theory has become a latter-day Holy Grail. We have a Lagrangian with one parameter, which would be fixed by experiment. You would then have a TOE. You could, in principle, explain all possible phenomena. Particle physics tells us that there are a huge number of elementary particles, which can be split into two categories: matter and force-carriers. The set of particles that define matter is composed of six quarks: u, d, s, c, b, t (up, down, strange, charm, bottom, top), while the force carriers are the photon, the electroweak bosons, Z, $W^{\pm}$, the graviton g, and eight gluons responsible for the strong force, and the recently discovered Higgs boson. So, in particle physics, we have a Lagrangian, which sums over all particle types and distinguishes between matter and force-carriers in some way. If we had a TOE, all the particles and forces should be unified in some way so that we could write down a Lagrangian for a "master entity," and the particles mentioned would then just be different manifestations of this underlying entity. In string theory, the underlying entity is the string, and different excitations of the string represent different particles. Furthermore, unification of the four fundamental forces is also built into the theory.

In the world we perceive, there are three familiar dimensions of space: height, width, and depth. Einstein's general theory of relativity treats time as a dimension on a par with the three spatial dimensions. In general relativity, space and time are not modeled as separate entities but are instead unified to a four-dimensional space-time; three spatial dimensions and one time dimension. In this framework, the phenomenon of gravity is viewed as a consequence of the geometry of space-time. In spite of the fact that the Universe is well described by four-dimensional space-time, there are several reasons why physicists consider theories in other dimensions. In some cases, by modeling space-time in a different number of dimensions, a theory becomes more tractable mathematically, and one can perform calculations to gain insights more readily. There are also situations where theories in two or three space-time dimensions are useful for describing phenomena in condensed matter physics. Finally, there exist scenarios in which there could actually be more than four dimensions of space-time, which have nonetheless managed to escape detection.[28]

However, one should keep in mind that string theory is in some sense only in its infancy, and, as such, is nowhere near being able to answer any of the questions we would wish it to answer; especially regarding what happens at singularities. There are those who believe that, in the end, string theory will either have nothing to do with Nature, or will never be testable, and as such will be relegated to being a mathematical plaything, or a new branch of philosophy.

11.3 REALITY

At present, there is no candidate TOE that includes the Standard Model of particle physics and general relativity. For example, no candidate theory is able to calculate the Fine Structure constant ($\alpha = 1/137.035\ 999\ 084(21)$), or the mass of the electron ($m_e = 9.109\ 383\ 56(11) \times 10^{-31}$ kg), both of which are known very precisely from experiment (as can be seen by the error limits). We know these fundamental constants with extraordinary precision from experiment, and achieving that level of agreement with a theory, would be a significant test on the validity of any theory. Most particle physicists expect that the outcome of the ongoing experiments, the search for new particles at the large particle accelerators and for dark matter, are needed in order to provide further input for a TOE. However, there are many who take the view that the search for the TOE is a waste of time and resources. In fact, no better than the futile search for the Perfect Language by Dante and the Kabbalists, or Isaac Newton and Gottfried Leibniz.

Kurt Friedrich Gödel (1906–1978) was an Austrian, and later American, logician, mathematician, and philosopher; he is considered to be one of the most significant of logicians. His incompleteness theorems define modern views of mathematical logic; they demonstrate the inherent limitations of every formal axiomatic system capable of modeling basic arithmetic. These results from 1931 are important both in mathematical logic and in the philosophy of mathematics. The theorems are widely, but not universally, interpreted as showing that David Hilbert's 1900 program to find a complete and consistent set of axioms for all physics, the TOE is impossible.

A number of scholars claim that Gödel's incompleteness theorem suggests that any attempt to construct a TOE is bound to fail. Gödel's theorem, informally stated, asserts that any formal theory expressive enough for elementary arithmetical facts to be expressed, and strong enough for them to be proved is either inconsistent (both a statement and its denial can be derived from its axioms) or incomplete, in the sense that there is a true statement that can't be derived in the formal theory. Stephen Hawking was originally a believer in the TOE but, after considering Gödel's theorem, concluded that an ultimate theory is not possible: *"Some people will be very disappointed if*

---

[28] There is a wonderfully imaginative use of such compactification and hidden dimensions in the science fiction of Liu Cixin; the technology of the Trisolarians and other alien species in *The Three-body Problem* of 2007, and its two sequels, *The Dark Forest* (2008) and *Death's End* (2010).

there is not an ultimate theory that can be formulated as a finite number of principles. I used to belong to that camp, but I have changed my mind."

Other physicists have argued against this view, pointing out that Gödel's theorems are irrelevant for computational physics; that is, purely model-based theoretical physics. Analogously, it may (or may not) be possible to completely state the underlying rules of physics with a finite number of well-defined laws, but there is little doubt that there are questions about the behavior of physical systems which are formally undecidable on the basis of those underlying laws. Whereas there may or there may not be an underlying philosophical/theoretical reason why a TOE may or may not exist, there is also a more prosaic argument about experimental uncertainty that clouds the search. How do you know if you have found a more fundamental model of physical reality, a higher synthesis of the Laws of Nature?

To date, no physical theory is held to be precisely accurate. Physics proceeds by a series of successive approximations, allowing more and more accurate predictions and measurements of an ever-wider range of phenomena; see Table 11.1, which summarizes the evolution of the precision of the measured values of the speed of light, c. Some physicists believe that it is therefore a mistake to confuse theoretical models with the true nature of reality, and hold that the series of approximations will never terminate in an "absolute truth." Einstein himself expressed this view on several occasions. Following this view, we may reasonably hope for a TOE which self-consistently incorporates all currently known forces, but we should not expect it to be the final answer.

A motive for seeking a TOE apart from the pure intellectual satisfaction of completing a centuries-long quest, is that prior examples of unification have predicted new phenomena, some of which (for example, electrical generators and technology) have proved of great practical importance to our civilization.

11.4 FURTHER READING

The Elegant Universe: Superstrings, Hidden Dimensions, and the Quest for the Ultimate Theory (2003); Brian Greene; New York: W.W. Norton & Company.

The Road to Reality: A Complete Guide to the Laws of the Universe (2005); Roger Penrose; Knopf.

The Trouble with Physics: The Rise of String Theory, the Fall of a Science, and What Comes Next (2006); Lee Smolin; New York: Houghton Mifflin Co.

Not Even Wrong: The Failure of String Theory And the Search for Unity in Physical Law (2006); Peter Woit; London, Jonathan Cape.

Table 11.1: The evolution of the experimental value of the speed of light, c. We see how the value of this constant of Nature converged to its present day accepted value, but the size of the uncertainty associated with the measurements also fell with time and the increasing precision of the measurements. Today, the value of c is fixed by the definition of the meter in the modern Quantum-SI; hence, the present value of c is exact, and without error. But this "fixing" of the value of a constant of Nature has implications on our evolving understanding of other aspects of Nature. All Nature is interconnected (see Figure 9.2), and we should be careful in formally constraining some of those connections.

Year	Method	Value of c
1675	Astronomical observations of the moons of Jupiter	220,000 km/s
1729	Studies of optics (parallax)	301.000 km/s
1849	Optics (interference)	315,000 km/s
1862	Optics (rotating mirror)	298,000 ± 500 km/s
1907	Electromagnetism	299,710 ± 30 km/s
1926	Optics (interferometry) due to Albert Michelson	299,796 ± 4 km/s
1950	Electromagnetics (masers)	299,792.5 ± 3.0 km/s
1958	Radio interferometry	299,792.50 ± 0.10 km/s
1972	Laser interferometry	299,792.4562 ± 0.0011 km/s
1983	c fixed with the new definition of the meter	299,792.458 km/s (Exact, and so without error.)

CHAPTER 12

Changing the Paradigm: From Long Lists to Short Explanations

One had to be a Newton to notice that the Moon is falling, when everyone else sees it doesn't fall.

Paul Valéry (1871–1945)

So how did we go from the earliest stage of the creation of science, that is, the creation of long lists, to looking for an underlying principle to explain all the observations and information contained in those long lists?

12.1 THE GREAT PARADIGM SHIFT IN BIOLOGY

In many ways, modern physics, or if you will modern physical sciences, is an unstable structure. While parts of the edifice of physics are solid enough, and have been around for centuries; representing a coherent story, there are other parts of the edifice that have been added in an *ad-hoc* manner. Quantum mechanics enables one to calculate many measureable atomic properties, and when the theoretical or calculated quantity is compared to the measured quantity, we have exceptional agreement, and we say that the theory must therefore be true. This is, after all, the basis of the scientific method. Yet quantum mechanics does not fit in at all with astrophysics. These two areas of physics each apply to very different length scales—from galaxies to quarks. Consequently, we have at present two very different and successful, in their own domains, models of reality, but they do not come together. We have yet to construct, or find the Theory of Everything (see Chapter 11).

When you look at physics today, you are looking at the state of biology in the early 1950s; that is, before the discovery of the structure of deoxyribonucleic acid or DNA, and the explanation of how this molecule and its self-replication explains evolution on earth. Before the early 1950s, we knew about the patterns of inheritance of characteristics such as eye color and hair color in animals, the breeding of horses and greyhounds, and the color of flowers in pea plants; and it was suspected that this strange mechanism of inheritance had something to do with the complex molecules found in the nuclei of all living cells. These large, complex molecules, which became known as nucleic acids, were investigated and found to be long polymers made up of a handful of smallish molecules. And so the race was on to try and determine the structure of the nucleic acid polymer to see if it could tell us something about inheritance and genetics.

The problem was solved in 1953. Francis Crick (1916–2004) and James Watson (born 1928) determined the double-helix structure of the major component of nucleic acid, deoxyribonucleic acid (DNA). Not only did they determine the position of the atoms within the double-helix, but they showed that when the DNA molecule divides, at the same time as the cell divides into two daughter cells, the two (helical) strands of the original DNA molecule unwind, and each strand assembles a new partner strand from small molecules available in the surrounding cell-fluid. And it is the rules that govern this assembling of a new stand of the double-helix DNA, based on the chemical structure of the original strand that allows physical characteristics to be passed from one generation to the next. Two physicists, using the technique of x-ray crystallography, had demonstrated that the whole of biology could be re-interpreted from the view-point of the three-dimensional structure of the complex molecules found in the cells of every living organism. What was more, the work of these two young crystallographers demonstrated that if a mistake were made in building the daughter-DNA molecule; that is, in assembling the new DNA strand/molecule to be fitted into the daughter nucleus then there was the possibility of a mutation, or a change in the blue-print of life of that organism. Generating the possibility that the next generation of that organism would be ever so slightly different from their parents. At a stroke, the mechanism of Darwin's theory of evolution by natural selection was discovered, and the pseudo-science of eugenics overturned.

The paradigm of biology had changed. A new unity had been achieved out of a synthesis of diversity. Before the 1950s, biology consisted in learning the contents of some very long lists contained in a vast array of books. With reference to what was said in Chapter 1, when it comes to retrieving data and information we had moved from Homer's *Catalogue of Ships* to the database of swift-footed Achilles. After the 1950s, biology consisted in looking at the observations contained in those long lists of book and re-interpreting them in terms of the structure and replication of DNA. The discovery of the structure of DNA, and how it replicated itself, revolutionized every aspect of biology and medicine. It opened up the possibility of designing life, and of significantly extending life-spans. The list of how the discovery of Crick and Watson will change our world is practically endless. This century will be the century of genetic engineering and genetic medicine, just as the last century was the century of the physical sciences.

12.2 ELECTROMAGNETISM

When we use a mobile phone, listen to the radio, use a remote control, or heat food in the microwave, few are aware that it was the great Scottish physicist/mathematician, James Clerk Maxwell (1831–1879), who was responsible for making these technologies possible. In 1865, Maxwell published an article entitled "A Dynamical Theory of the Electromagnetic Field," where he stated: *"It seems we have strong reason to conclude that light itself (including radiant heat, and other radiations if any) is an electromagnetic disturbance in the form of waves propagated through the electromagnetic field according to electromagnetic laws.*" These ideas of Maxwell ushered in the great synthesis of electromagnetism. This paradigm change was a development of the metric system in Revolutionary France (April, 1795). In the same way that the study of heat and energy had not reached a sufficiently mature stage to allow the *savants* who formulated the metric system to propose a base unit for temperature in 1795, the study of electricity was at an even more immature stage. Indeed, the study of electricity and magnetic phenomena in the 1790s had more in common with parlor tricks than laboratory investigations. The two names associated with the earliest investigations of the nature of electricity are the pious, conservative, Italian medical doctor Luigi Alyisio Galvani (1737–1798), and another Italian, the natural philosopher Alessandro Giuseppe Antonio Anastasio Volta (1745–1827).

In 1791, Galvani famously discovered that the leg muscles of dead frogs twitched when they came in contact with an electrical spark. According to popular versions of the story, Galvani was dissecting a frog at a table where he had previously been investigating discharges of static electricity. Galvani's assistant touched an exposed sciatic nerve of a dead frog with a metal scalpel, which had picked up a residual static (electrical) charge. At that moment, the two men saw the leg of the dead frog kick as if it were alive. Such laboratory-based observations made Galvani the first to consider the possible relationship between electricity and animation; that is, the creation of life, and the possibility of the re-animation of dead tissue. Indeed, Luigi Galvani used the term "animal electricity" to describe the force that animated the muscles of the dead frog. Along with many of his contemporaries, he regarded the activation of the supposedly dead muscles as being generated by an electrical fluid carried by the still functioning nerves of the frog to the inanimate muscles. Given his background, Galvani naturally assumed he had discovered something of the animating or vital force that was implanted in all creatures by their Creator. However, not everyone agreed with this conclusion. In particular, Alessandro Volta thought that the term "animal electricity" had a suggestion of superstition and magic, and that it was not an explanation of the dramatic phenomenon observed repeatedly by Galvani and co-workers. For his part, Galvani held that natural philosophers like Volta had no place in moving from the laboratory into God's realm of vitalism and the nature of life itself. The argument between Galvani and Volta was a microcosm of the larger debate about the place of the Divine in Nature which was animating the European Enlightenment. Galvani spent years repeating his experiment on dead animals, and discovered that you did not need a traditional source of static electricity to cause the dead muscle tissue to twitch. A combination of two wires of different metals, for example, copper and zinc was sufficient, but Galvani could not explain these observations.

The phenomenon observed by Galvani was subsequently named "galvanism," on the suggestion of his sometime intellectual adversary, Volta. Today, the term galvanism is used only to describe someone who suddenly becomes excited, and it is likely that most people who use this word have no idea of its origin. Although at the beginning of the 19th century, the observations of Galvani were the source of much discussion, most famously in the novel *Frankenstein, or, The Modern Prometheus* by Mary Shelley, which describes further investigations into the principles of animation and vitalism.

Alessandro Volta was more of a scientist than Galvani. In the late 1770s, Volta had studied the chemistry of gases, and was the first person to investigate the origin and chemical composition of natural gas, or methane. However, it is for his investigations into the nature of electricity that Volta is most famous, in particular, for a systematic investigation of electrical capacitance. He developed separate means of investigating both the electrical potential applied to the two plates of the capacitor, and the charge residing on the plates. Volta discovered that for a given pair of plates, the potential and the charge are proportional. This relationship is called Volta's Law of Capacitance, and to honor his fundamental work on electrostatics the unit of electrical potential is named the volt.

Alessandro Volta realized, from his own studies of Galvani's observations that the frog's leg merely served as both a conductor of electricity (the fluid in the dead muscle tissue is what today we would term an electrolyte) and a detector of the presence of a flowing electric current; all of which mimicked an instantaneous animation. Indeed, Volta realized that the two different metals (the electrodes) used by Galvani, inserted into the fluid of the frog's leg formed an electrical circuit. Volta replaced the frog's leg by paper saturated with another conducting electrolyte, e.g. salt solution, and detected a flow of electricity. In this way he invented the electrochemical cell, the forerunner of all chemical batteries.

Luigi Galvani never perceived of electricity as being separable from biology. He always believed that animal electricity came from the muscle of the animal. Volta, on the other hand, reasoned that animal electricity was merely a physical phenomenon external to the dead frog, an electric current coming from the metals, which formed an electrochemical cell or battery (for example, zinc and copper), mediated by the fluid in the muscle tissue. There was no reanimation of dead tissue, merely a flow of electrical current from one electrode to the other electrode through the physiological fluid (the electrolyte) in the muscle of that poor dead frog. But Galvani's ideas did give literature, and the cinema Dr Frankenstein and his splendid creature.

In the early 19th century, electricity, magnetism, and optics were three independent disciplines. However, the situation changed thanks to one invention and two discoveries. The invention was the electrical battery, a continuous source of electrical current created by Alessandro Volta in about 1800. The two discoveries were: (1) the demonstration of magnetic effects caused by the flow of electrical currents, observed by the Danish chemist and physicist Hans Christian Ørsted (1777–1851) and by the French mathematician and one of the creators of the Metric system, André-Marie Ampère (1775–1836) in 1820; and (2), the 1831 discovery by the British chemist and natural philosopher Michael Faraday (1791–1867) of the generation of electrical currents from magnetic fields, that is, electromagnetic induction. In September 1820, Ampére presented his results to the *Académie des sciences*: "*mutual action between currents without the intervention of any magnet*"; that is, two parallel electrical currents attract, or repel each other depending on their polarity, as do permanent magnets. In 1826, he published *Theory of Electrodynamic Phenomena, Uniquely Deduced from Experience*, whereby he claimed that "*magnetism is merely electricity in motion*" and that magnetic phenomena depend only on the existence and motion of electrical charges (see the lines of force emanating from a bar magnet, which also represents the magnetic field generated by an electric current in Figure 12.1), thereby setting the stage for Faraday's experiments.

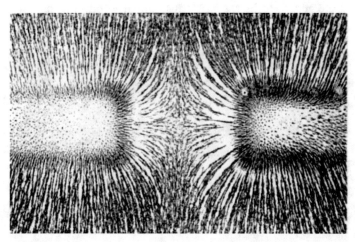

Figure 12.1: The invisible repulsive lines of force emanating from similar poles of two bar magnets; visualized by the use of iron fillings. Image from: https://commons.wikimedia.org/wiki/File:Magnetic_field_of_bar_magnets_repelling.png.

The three contributions mentioned above form the basis of modern electromagnetism, but required the insight of the Scot, James Clerk Maxwell to form a coherent single theory. Before Maxwell, electromagnetism still consisted of long lists of observations of supposedly disparate phenomena; Maxwell demonstrated the single underlying causation. Such a synthesis represents the most profound transformation of the fundamentals of physics since Newton, and is one of the greatest of scientific achievements, unifying electrical and magnetic phenomena, and enabling the development of the theory of electromagnetic waves, including light.

James Clerk Maxwell published his major work, *A Treatise on Electricity and Magnetism* in 1873; a first step on the great journey to the Theory of Everything (see Figure 11.2). Here, Maxwell rationalized and unified all the then known phenomena involving electricity and magnetism. When we come to consider how matter interacts with light; that is, with an oscillating, or time-varying electromagnetic field, we have to consider the other great contribution to the final synthesis of electromagnetism made by Maxwell, who, between 1861 and 1862, published a set of equations relating electricity and magnetism and demonstrated that light is another electromagnetic phenomenon. Classically, light scattering arises through secondary radiation from oscillating dipoles induced by the incident electromagnetic wave. The simplest case occurs when the scattering medium is a gas, composed of randomly distributed molecules of dimensions that are small compared to the wavelength of the light.[29] For a random distribution, the phase relationships between waves scattered from different molecules are uncorrelated in all but the forward direction, so that the total scattered intensity can be calculated directly as the sum of contributions from each molecule; thereby permitting study of the properties of individual scattering molecules. Figure 12.2 demonstrates the dramatic colors that are generated by the scattering, and absorption of the light coming from the Sun, by the molecules in the atmosphere.

Figure 12.2: The blue color of the sky is caused by light scattered by atmospheric gas molecules (N_2, O_2, H_2O, and CO_2), and not by absorption. These molecules being much smaller than the wavelengths of visible light. The red color at sunset and sunrise (sunrise in Montpellier, France, in October 2019, in this photograph) comes from absorption, because at sunrise and sunset the Sun is low in the sky and so the sunlight is passing through the thickest section of atmosphere (that is, the path-length is at its longest). This absorption of blue light, leaving the red/pink color is due to molecules other than the normal components of air; that is, pollutants or dust. The grey/white color of clouds is caused by light scattered by water droplets, which are of a comparable size to the wavelengths of the incident visible light. The darker the color of the clouds, the larger are the water droplets, as liquid water does have a weak (electric dipole forbidden) absorption in the visible.

---

[29] The molecules comprising air (principally, N_2 and O_2) have a "diameter" of about 1 Å; that is, 1×10^{-10} m. On the other hand, the wavelength of visible light; that is, the spacing between two successive maxima of the oscillating electromagnetic wave is about 5,100 Å (in the green where our eyes have evolved to be the most sensitive).

In the next two chapters, we will consider how the classification of information about life, and about natural phenomena is accomplished, and how this scientific classification assists scientists in comprehending Nature, and in developing theories to explain the world around us. First we will consider biology. As pointed out above, the paradigms of biology changed with the discovery of the structure and function of DNA, and the interpretation of evolution at a molecular level in terms of the hydrogen-bonding between the four nucleoside bases (the monomers) that compose the large (long) polymeric DNA molecule: adenine, cytosine, guanine, and thymine. The diversity and evolution of life on earth arises from the coupling of these four smallish molecules, when they are bound into the DNA polymer. It is generally held that evolution is the most powerful and comprehensive idea ever formulated.

12.3 FURTHER READING

1   *The Double Helix* (1968); James D. Watson; New York, Touchstone.

2   *The Molecule as Meme* (2018); Jeffrey H. Williams; San Rafael, CA:Morgan & Claypool.

CHAPTER 13

The Classification of the Living and the Dead

Nothing in biology makes sense except in the light of evolution.

Theodosius Grygorovych Dobzhansky, (1900–1975)

There are millions of species of organisms, both plants and animals living on this Earth. In addition, there are many millions of species preserved in the fossil record. These are plants and animals that lived once upon a time, and as a consequence of climate change, natural selection, and rogue large meteorites have become extinct. Given the scientist's desire, and need for classification and list making, how is it possible to keep track of everything that is alive, or has ever been alive? This is not a dull academic question, as if we are to understand how we (*Homo sapiens*, or "thinking man") evolved from less-advanced creatures; we must be able to locate ourselves, and our ancestors in the Great Scheme of Life on this planet. How then do we name and organize all of the long lists of the living and the dead, without getting confused, and without our imperfect memories leaving large embarrassing lacunae in our models of the evolutionary story of life on this planet?

The answer to this question is straightforward; we use a complex, but elegant system of classification developed in the 18th century by Carolus Linnaeus (1707-1778).[30] Linnaeus was a Swedish botanist, physician, and zoologist who formalized a binomial nomenclature of organisms; and we still use this system of naming and classifying organisms. Indeed, Linnaeus is known as the "father of modern taxonomy" (see Figure 13.1). Linnaeus was a product of, and became a towering figure in, the European Enlightenment. However, he is less well known,or remembered today than many of his scientific and philosophical contemporaries, particularly, those from France, Germany, and Scotland.

---

[30] Many of his writings were in Latin, and his name Carl von Linné, is rendered in Latin as Carolus Linnaeus. In addition, we speak of the Linnaean system of classification (from the Latin form of his name), but the Linnean Society of London is named from the original Swedish spelling of the name. The Linnean Society of London are the custodians of Linnaeus' specimen collection, and are the UK's learned society responsible for taxonomy. It was in the rooms of the Linnean Society that the papers of Charles Darwin and Alfred Russel Wallace on evolution by natural selection were presented on July 1, 1858.

Figure 13.1: Painting of 1737 by Martin Hoffman showing Carl von Linné (Linnaeus) in his field-costume; the traditional dress of the Sami people of Lapland. In his hand is the plant that Jan Frederik Gronovius named after him. *Linnaea borealis* is a species of flowering plant in the family *Caprifoliaceae* (the honeysuckle family). Until relatively recently, it was the only species in the genus *Linnaea*. It is a boreal to subarctic woodland subshrub; hence the specific name, and is commonly known as twinflower. This plant was a favorite of Carl Linnaeus, founder of the modern system of binomial nomenclature, for whom the genus was named (image from: https://en.wikipedia.org/wiki/Carl_Linnaeus#/media/File:Naturalis_Biodiversity_Center_-_Martin_Hoffman_-_Carl_von_Linné_(Linnaeus)_in_his_Lapland_costume_-_painting.jpg).

The European Enlightenment was the 18th century up to, but not including, the French Revolution; the great socio-political event that was the product of the Enlightenment. This was a period of intense philosophical and scientific investigation. Toward the end of the Enlightenment, in 1784, the German philosopher and mathematician Immanuel Kant (1724–1804) published his celebrated essay, "Answering the Question: What is the Enlightenment?," where he told his readers that the Enlightenment was man's emergence from a self-imposed immaturity that had led to his incapacity to use his understanding to explain Nature without guidance from another (higher) being. Kant told his readers that they had to be courageous to understand Nature, they had to *"dare to know"* (in Latin, *sapere aude!*). That man should investigate the world around him, and pursue his investigation to the limit of technology; and then use his mind to imagine what might happen beyond his technical limitations. Although best known today for his work in ethics and metaphysics, Kant made significant contributions to other disciplines, especially, mathematical physics. In 1754, he was awarded the Berlin Academy Prize for his prediction of the inevitable slowing down of the Earth's rotation.

In his early mathematical studies, Kant pointed out that due to the gravitational attraction between the Earth and the Moon, the frictional resistance of the motion of the oceans on the Earth's surface must lead to a slow decrease in the Earth's speed of rotation. Energy was being generated as friction between the moving slabs of water, and the rotating solid Earth that supported and carried the oceans. Kant knew that energy was conserved; it cannot be made or be lost, so he reasoned that the frictional interaction of the oceans with the massive rotating crust had to come from somewhere, and Kant further reasoned that it was being taken from the speed of rotation of the Earth; that is, the Earth's angular momentum. The position of the Moon relative to the Earth is determined by the sum of the attractive forces derived from the masses of the Earth and Moon and the much larger, but more distant, mass of the Sun. That is, the overall gravitational attraction of these three bodies. Through this mutual gravitational force, the Earth holds the Moon to itself and the Moon generates the tides seen on Earth; tides that are the origin of the inevitable slowing of the Earth's rotation (2.3 milliseconds/century). This discovery attracted little or no attention until about 1840, when the concept of energy began to be widely and more fully comprehended. Yet, the slowing of the Earth is the reason that time-metrologists still insert "leap seconds" into the year to maintain coherence between Greenwich Mean Time (a measurement of when the sun rises at Greenwich) and atomic time (a measure of frequency of an excitation within the hydrogen atom).

These early mathematical investigations had taught Kant that all the sciences are linked; that Nature is a seamless whole (see Figure 9.2). He looked at the energy of interaction of the oceans with the Earth's rotating solid surface, then using mathematical logic he saw that the Earth is a complex but self-contained, and self-regulating system. He dared to use mathematics to follow, to its logical conclusion something that no one had previously considered: that the Earth is inevitably slowing down through the generation of frictional energy. A mechanical energy derived from the relative motion of the oceans to the Earth. Kant clearly demonstrated that one should look at Nature holistically; in much the same way that the Ancient Chinese Taoist sages had looked at the world around them. And in so doing, Kant came to a startling conclusion. What Kant and others achieved in the physical science, by building on the work and ideas of Isaac Newton, the Swedish botanist Carl Linnaeus attempted in the monumental task of trying to devise a system of classification capable of containing not only every living organism, but every organism that had ever lived. And in so doing provide a wealth of information about the interconnectedness of those living and extinct organisms; connections which proved invaluable after the appearance of Darwin's explanation of the evolution of species in the mid-19th century.

Carl von Linné (Carl Linnaeus) was born in the countryside of Småland in southern Sweden. He received most of his higher education at Uppsala University and began giving lectures in botany in 1730. He lived abroad between 1735 and 1738, where he studied and published the first edition of his systematic classification of Nature, *Systema Naturae,* in the Netherlands. He returned to Sweden, where he became professor of medicine and botany at Uppsala. In the 1740s, he was sent on several journeys through Sweden to find and classify plants and animals. In the 1750s and 1760s, he continued to collect and classify animals, plants, and minerals, while publishing several further volumes. At the end of his life, he was one of the most acclaimed scientists in Europe. Philosopher Jean-Jacques Rousseau sent him the message: *"Tell him I know no greater man on earth."* Johann Wolfgang von Goethe wrote: *"With the exception of Shakespeare and Spinoza, I know no one among the no longer living who has influenced me more strongly."* Swedish author August Strindberg wrote: *"Linnaeus was in reality a poet who happened to become a naturalist."* Linnaeus has also been called *Princeps botanicorum* (Prince of Botanists), and is considered as one of the founders of modern ecology. But it is as a taxonomist[31] that he is remembered today.

13.1   A HIERARCHICAL SYSTEM OF CLASSIFICATION

During his lifetime, Linnaeus collected around 40,000 specimens of plants, animals, and shells. He believed it was important to have a standard way of grouping and naming species. So in 1735, he published the first edition of *Systema Naturae* (*The System of Nature*), which was a small pamphlet explaining his new system of the classification of Nature. He continued to publish further editions of *Systema Naturae* that included increasing numbers of named species. In total, Linnaeus named 4,400 animal species and 7,700 plant species using his binomial system of nomenclature. The tenth edition of *Systema Naturae* was published in 1758 and is considered the most important edition; its full title in English is *System of Nature through the Three Kingdoms of Nature, According to Classes, Orders, Genera, and Species, with Characters, Differences, Synonyms, Places.* In *Systema Naturae*, Linnaeus classified Nature into a hierarchy. In this, Linnaeus was following the classification of John Wilkins in his attempt to create a new philosophical universal language in 1688 (see Page 57). Linnaeus proposed that there were three broad groups, called kingdoms, into which the whole of Nature could be fitted. These kingdoms were animals and plants; he originally attempted to classify minerals within the same hierarchy, but this did not work. He divided each of these kingdoms into classes; classes were divided into orders. These were further divided into genera (genus is the singular) and then into species. We still use this system today, but we have made some changes.

Today, we only use this system to classify living things, or things that were once alive. Also, we have added a few additional levels in the hierarchy. The broadest level of life is now a domain. All living things fit into only three domains: *Archaea* (single-celled microorganisms), *Bacteria* (prokaryotic microorganisms) and *Eukarya* (organisms whose cells have a nucleus enclosed within membranes, unlike prokaryotes (*Bacteria* and *Archaea*), which have no membrane-bound organelles). Within each of these domains there are kingdoms, for example, *Eukarya* includes the Kingdoms: *Animalia*, *Fungi*, *Plantae* (plants). Each kingdom contains phyla (the singular is

---

[31] Taxonomy is the part of science that focuses on naming and classifying, or grouping organisms. Carolus Linnaeus developed a way of naming and organizing species that we still use today, and which is still expanding as new species of living and extinct organisms are discovered.

phylum), followed by class, order, family, genus, and species. Each level of classification is called a taxon (the plural is taxa).

According to this system, the tree of life consists of three domains: *Archaea, Bacteria,* and *Eukarya.* The first two are all prokaryotic microorganisms, or single-celled organisms whose cells have no nucleus. All life that is made up of cells containing a nucleus and membrane-bound organelles, and multicellular organisms, is included in the *Eukarya.* Kingdom is the second highest taxonomic rank, just below domain. Kingdoms are divided into smaller groups called phyla (except the K. *Plantae,* which has "Divisions").[32]

Some recent classifications based on modern cladistics[33] have explicitly abandoned the term kingdom, noting that the traditional kingdoms are not monophyletic; that is, do not consist of all the descendants of a common ancestor. Depending on definitions, the animal kingdom *Animalia* or *Metazoa* contains approximately 35 phyla, the kingdom *Plantae* contains about 14, and the kingdom *Fungi* contains about 8 phyla. The total numbers of species in these phyla are estimates; figures from different authors vary wildly, not least because some are based on described species, some on extrapolations to numbers of undescribed species. And then there is the problem of estimating, from the fossil record the number of species of a particular phylum that existed in the distant past. For instance, around 25,000–27,000 species of nematodes have been described, while published estimates of the total number of nematode species include 10,000–20,000; 500,000; 10 million; and 100 million.

As mentioned above, animals, fungi, and plants are arranged into various groupings to assist with the classification of the great variety of life on Earth. Extinct organisms are treated in the same way as extant organisms (those which are still around today). All life on Earth belongs to one of three kingdoms: *Animalia, Plantae,* and another kingdom for the *Fungi.* Kingdoms are further sub-divided into other categories, organizing creatures, plants, and fungi in such a way that common features lead to organisms being associated together until an individual species is defined; that is, until we arrive at the specific. This is the accepted method of classifying life (although cladistics has added a new dimension, or two), the principles of this form of classification were laid down by Carolus Linnaeus in the 18th century. To have achieved this level of hierarchical classification would have been impressive enough, but Linnaeus also demonstrated that by judicious choice of

[32] Traditionally, some textbooks from the U.S. used a system of six kingdoms (*Animalia, Plantae, Fungi, Protista* (any eukaryotic organism (one with cells containing a nucleus) that is not an animal, plant or fungus), *Archaea/Archaebacteria,* and *Bacteria/Eubacteria*) while textbooks in countries such as Great Britain, India, Greece, Australia, Latin America used five kingdoms (*Animalia, Plantae, Fungi, Protista,* and *Monera* (a kingdom that contains unicellular organisms with a prokaryotic cell organization, having no nuclear membrane such as bacteria).

[33] A method of classification of animals and plants that seeks to identify and take account of only those shared characteristics which can be deduced to have originated in the common ancestor of a group of species during evolution, not those arising by convergence.

names, for the various taxa, one could input into the classification a great deal of useful morphologic information; see Table 13.1.

Category			
	(Taxon)	(Taxon)	(Taxon)
Domain	*Eukarya*	*Eukarya*	*Eukarya*
Kingdom	*Animalia*	*Animalia*	*Animalia*
Phylum	*Chordata*	*Chordata*	*Chordata*
Class	*Mammalia*	*Mammalia*	*Reptilia*
Order	*Primates*	*Carnivora*	*Saurischia (Theropoda)*
Family	*Hominidae*	*Felidae*	*Tyrannosauroidea*
Genus	*Homo*	*Felis*	*Tyrannosaurus*
Species	*H. sapiens*	*F. catus*	*T. rex*

Table 13.1: Classification of man, an animal that lives with man, and an extinct carnivorous dinosaur

The specimen of *T. rex* shown here is in the Field Museum in Chicago, IL, USA, and is affectionately known as Sue. There are many photos of Sue all over the Internet. The cat lives in south London.

We see from Table 13.1 that a human, a cat, and an extinct meat-eating dinosaur are all in the same domain, kingdom, and phylum. All three of us are, or were, multi-celled animals with backbones (the phylum *Chordata*: having a spine or backbone). It is after this level in the Great Hierarchy of Life that there is a divergence. The human and the cat are mammals (*Mammalia*), but the dinosaur was a reptile (*Reptilia*). Thereafter, the man and the cat diverge; we are *Primates*, and the cat is a *Carnivore*.

The categories of classification from kingdom down to the species level is referred to as a taxonomic hierarchy. Organisms should be classified to reflect evolutionary relationships, with each taxon representing organisms that share a common ancestor; this is, the Tree of Life model of taxonomy. It can be seen from Table 13.1 that the author, his neighbor's cat, and a large meat-eating dinosaur all share the same domain, kingdom, and phylum. It is only there after that we start to separate; that is, the tree of life branches for us. The author and the cat, both being mammals, lie on a different line of descent from dinosaur *T. rex*, which derived from the chordates via reptiles. To be absolutely correct the name of all taxa should begin with a capital letter, except for the individual species name which should always begin with a lowercase letter.[34] The scientific name for a particular organism consists of two Latin or Latinized words that are always the genus followed by the species classification. Hence the term binomial classification. Such a standard system of nomenclature ensures that scientists from around the world can communicate effectively when describing the characteristics of an individual organism, or even of a single fossil. The genus, such as *Tyrannosaurus*, can be used on its own but the species name; that is, *rex* without the genus associated with it has no meaning, as some species names may be used many times for organisms in different genera.

It is at the level of order that the three creatures illustrated in Table 13.1 separate fully into primate, carnivore, and dinosaur. At the next level down in the hierarchy, family, we refer to *Hominidae* (whose members are known as great apes or *hominids*, are a taxonomic family of primates that includes eight extant species in four genera: *Pongo*, the Bornean, Sumatran, and Tapanuli orangutan; *Gorilla*, the eastern and western gorilla; *Pan*, the common chimpanzee and the bonobo; and *Homo*, which includes modern humans and their extinct relatives (e.g., the Neanderthal; *Homo neanderthalensis*), and ancestors, such as *Homo erectus*); for the cats, *Felidae* (a family of mammals in the order *Carnivora*, colloquially referred to as cats. A member of this family is also called a *felid*. The *Felidae* species exhibit the most diverse fur pattern of all terrestrial carnivores.); and *Tyrannosauroidea* (meaning tyrant lizard forms is a group of coelurosaurian theropod dinosaurs that includes the family *Tyrannosauridae* as well as more basal relatives). The genus taxa in Table 13.1 become even more narrowly defined: *Homo* (from the Latin for human being, is the genus which emerged from the otherwise extinct genus *Australopithecus*, that encompasses the extant species *Homo sapiens* (modern thinking man), plus several extinct species classified as either ancestral to or closely related to modern humans (depending on a species), most notably *Homo erectus* and *Homo neanderthalensis*); *Felis* (genus of small and medium-sized cat species native to most of Africa and south of 60° latitude in Europe and Asia to Indochina). But it is with the specific taxon that we arrive at the specimens illustrated in Table 13.1. *Homo sapiens* (not, *Homo neanderthalensis*); *Felis catus* (not, *Panthera tigris*, the tiger) and *Tyrannosaurus rex*.[35]

In this way, all scientists, including palaeontologists work primarily with extinct species have a classification framework to use, and into which they may locate their particular specimen. That specimen will then be located in geological or evolutionary time, and in space; that is, in a particular habitat. However, there are additional conventions to consider when classifying and naming animals such as dinosaurs, (or indeed all organisms for that matter). If a dispute arises as to the

---

34 This is some of the dogma of science that has attached itself to the technical advances of science. Such dogma, or "the way things should be done, as decided by an international group of elderly scientists" is also encountered in chemical nomenclature (see Chapter 14) and in the use of the International System of Units (SI, abbreviated from the French *Système international d'unités*), the modern form of the metric system, and the most widely used system of measurement (see Chapter 8).

35 Only species are real. Everything else is interpretation; that is, arbitrary and often subject to re-classification. But a species name should be eternal (unless a precedent turns up).

naming of an organism (and they do; see Chapter 14 on the naming of the chemical elements) then it is convention for the earliest name, the first description to take precedence. For example, the nomenclature of early *hominids* (our direct ancestors) is a minefield (see comment below). But this system of priority publishing does permit standardization; in this way, the name *Brontosaurus* was replaced by its older synonym *Apatosaurus*. There have, however, been some notable exceptions to this, and *T. rex* is one of them.

In the late 19th century, many years before *T. rex* was named and described in 1905, the American palaeontologist, Edward Drinker Cope described two badly eroded fossil vertebrae as *Manospondylus gigas*; that is, "giant porous vertebra" in reference to the numerous openings for blood vessels he found in the fossilised bone. This strange honey-combed backbone was different to any known dinosaur fossils, and it was given this name. One of these bones has since been lost, however, the name stood and if scientific nomenclature was followed, as this bone is believed to represent a *Tyrannosaurus rex* then *T. rex*, the "Tyrant Lizard King" should be renamed *Manospondylus gigas* or "giant porous vertebrae," but that is not quite such an exciting name.

The debate as to the true name of *Tyrannosaurus rex* was brought to wider public attention when in 2000 a team from the Black Hills Institute of Geological Research, Hill City, South Dakota, U.S., www.bhigr.com , claimed they had found the original site where Cope had unearthed the weathered fossil bones described as *Manospondylus*. Fossils found on this site, presumably from the same specimen that Cope studied almost a century earlier, turned out to be *T. rex*, so *Tyrannosaurus rex* should have been renamed based on this evidence. The International Code of Zoological Nomenclature (ICZN) states that if further remains are found and these are identical to those of the earlier discovery then the earlier name and description should be used. This led to much consternation among scientists (and among Hollywood directors, as *Manospondylus* sounds significantly less cool than *Tyrannosaurus rex*). However, in 2000 the ICZN ruled that *T. rex* should stay, as the name had been cited in numerous works by many authors and the case of mistaken identity was more than 50 years old.

The creation of names, of naming things is an important and mysterious action, which is the origin of the attraction of the study of names, how to encapsulate in a word, or perhaps two words (as in the Linnaean system of biological nomenclature), the important characteristics of a newly discovered mineral, plant, animal or planet.[36]

---

[36] A small bird like the thrush has very different names in different countries, yet even if you know all those names, you would still know nothing about the bird. You would only know something about the people who have observed that bird, and what they called the bird. The thrush sings, it teaches its young to fly, and it flies great distances; distances so large that we are not sure how this small bird is able to navigate using the Earth's magnetic field as a guide. A true description of the thrush should provide some of this important information. The classification of the thrush: Kingdom: *Animalia*; Phylum: *Chordata*; Class: *Aves*; Order: *Passeriformes*; Suborder: *Passeri*; Family: *Turdidae*. Below this hierarchical level, the taxonomy of thrushes becomes complex because evolution is continually at work, leading to complex local characteristics.

13.2   A WARNING TO THE UNWARY

The Linnaean system of biological nomenclature is in two parts, allowing some personal individuality while retaining some information on sample characteristics. The first name refers to the genus and is given to a set of distinct anatomical characteristics of the organism. The second, specific name can take this morphological characterisation further; for example, the large ammonite (a member of the phylum *Mollusca* (molluscs) similar to the still extant *Nautilus*, but the ammonites became extinct along with the dinosaurs at the end of the Cretaceous Period about 70 million years ago)[37] found on Portland Isle, Dorset, UK, *Titanites giganteus*, is the largest representative of a genus of ammonite already well known for their size; see Figure 13.2. Although there are even larger ammonites in other genera.

Figure 13.2: Giant Ammonite of the genus *Titanities*, in an organic limestone (the rock consists largely of a matrix of shells) of Portland stone (Jurassic). Image from: http://www.southampton.ac.uk/~imw/portfoss.htm. Thanks to Dr. Ian West for permission to reproduce this image; he retains full copyright.

Alternatively, the second name in a Linnaean system of nomenclature may be the Latinized name of the person who first identified, or described the sample; the discoverer does not, however,

---

[37] Classification of Ammonites: kingdom: *Animalia*; phylum: *Mollusca*; class: *Cephalopoda*; subclass: *Ammonoidea*.

propose his or her own name—this is discretely left to colleagues and vice versa. Thinking up such names might seem a good way to relax after all the hard work of obtaining the specimen. However, any slackness or lack of care in determining the taxonomy of a new specimen can have serious repercussions; taxonomy is an important science as it explains and summarises the history of life on Earth. In 1995, an issue of *Nature* carried a scathing editorial condemnation of a research group who had, for whatever reason, ascribed a frozen corpse (affectionately named Ötzi, who died about 3345 BCE, and was found in an Austrian glacier) to a new species. The *Nature* editorial ran, "*With breath-taking abandon, Lubec et al. assign Ötzi to a new species,* Homo tirolensis. *No reason is given for this casual designation. Readers will look in vain for the careful systematic and diagnostic argument that such nomenclature requires,*" [1] Ötzi is as much a member of the species, *Homo sapiens* as is the author illustrated in Table 13.1. Such a colossal error in taxonomy (or, perhaps a lack of appreciation for the details of taxonomy) will terminate your scientific career faster than a large incoming meteorite.

13.3   THE LIMITS OF LINNAEAN CLASSIFICATION: TWO UNCLASSIFIABLE SPECIES FOUND OFF AUSTRALIA

As a final point, it is often the case that the limits of the Linnaean system of nomenclature are pushed by the discovery of new, exotic specimens. In this manner, the Linnaean system is continually being expanded, and refined. The following is taken from a report in the Guardian newspaper (www.theguardian.com/environment/2014/sep/04/two-unclassifiable-species-found-off-australian-coast) concerning a newly discovered organism. An organism that was so unusual that identifying the appropriate kingdom was problematic, and identifying the appropriate phylum almost impossible. The specimens displayed in Figure 13.3 were collected off the south-east coast of Australia in 1986. They were collected at water depths of 400 meters and 1,000 meters on the continental slope near Tasmania, using a sled that was dragged over the sea floor to collect bottom-dwelling organisms. The researchers were immediately struck by the unusual characteristics of the specimens collected.

When initially discovered, the organism's classification was difficult. The two specimens were assigned their own genus, *Dendrogramma*, and family, *Dendrogrammatidae* and the researchers even considered putting them in their own phylum. As they put it, however, "*we refrain from erecting such a high-level taxon for the time being, because new material is needed to resolve many pertinent outstanding questions.*" The lead scientist of the identification effort, Jørgen Olesen of the University of Copenhagen, suggested that they represent "*an early branch on the tree of life, with similarities to the 600 million-year-old extinct Ediacara fauna.*"

The genus name *Dendrogramma* derives from the two Greek words *déndron*, meaning tree-like, and *grámma* meaning drawing, mathematical figure. Alluding to the branching pattern of the digestive canals, see Figure 13.3, which resemble dendrograms; that is, branching diagrams frequently used by biologists to illustrate the evolutionary relationships among organisms. The specific name *enigmatica* of the type species refers to the mysterious nature of the organisms, while *discoides*—the species epithet of the second species—alludes to the disc-like shape of the animals.

Figure 13.3: Enigmatic specimens dredged up from the ocean depths. Preserved specimens of *Dendrogramma*. Images from: https://en.wikipedia.org/wiki/Dendrogramma#/media/File:Multiple_Dendrogramma.png.

13.4   FURTHER READING

The author is not a biologist, and is unfamiliar with texts used for teaching taxonomy; however, the author has found invaluable many of the articles available in the online encyclopaedia, Wikipedia (https://en.wikipedia.org/wiki/Wikipedia). In particular, the articles on Linnaean taxonomy (https://en.wikipedia.org/wiki/Linnaean_taxonomy) and taxonomy (https://en.wikipedia.org/wiki/Taxonomy_(biology)) were helpful, as were the links they contain.

1   *Nature*, 1995, 373, 176.

CHAPTER 14

Aspects of Chemical Nomenclature

What's in a name? that which we call a rose / By any other name would smell as sweet

Romeo and Juliet, Act 2, Scene 2; William Shakespeare (1564–1616)

The invention of modern biology began with the order introduced by Linnaeus, with his binomial nomenclature. In this way, biology became more that just long lists of organisms, fossils, and observations of morphology and behavior. Biology became a system, which was developed further by Charles Darwin, and finally rationalized by the work of Crick and Watson in the 1950s. At a stroke, Darwin's theory of evolution became a fact grounded in molecular physics. The myriad of books on everything from the breeding of horses and pigeons, to the genetics of pea plants and roses could be replaced by a new paradigm based on the interaction, and number of the hydrogen-bonds formed between the four DNA nucleoside bases. Biology and condensed matter physics were unified, and those long lists could be forgotten.

This same type of rationalization is underway in chemistry. There are about 20 million known chemicals, with new molecules being synthesised every week. How do you construct a systematic chemical nomenclature, which removes the appalling idea of having to memorise all, or even a smallish part of those individual trivial names? Well, the subject of chemical nomenclature is truly vast. Indeed, the International Union of Pure and Applied Chemistry (IUPAC) was created over a century ago to address this very problem. But sadly, not everyone in the chemistry community follows the rules of nomenclature laid down by the committees of experts in IUPAC (see https://iupac.org/what-we-do/nomenclature/ for details). Here we will consider the names of the chemical elements, which amply demonstrate the limitations of the present system of nomenclature. And of how the present system of nomenclature is not based purely on dispassionate scientific argument. Figure 14.1 and Table 14.1 give us a feeling for the arcane origins of chemistry, and of chemical nomenclature.

We are taught that science is above politics; that it, science is truly international, but is this really true? The more one questions, for example, the names selected today for newly discovered chemical elements, phenomena, and units the more one seems to see the world of politics and the individual intrude into the world of science. But then the selection of a name is not a trivial matter. Old taboos about not revealing one's name to a stranger, lest that stranger gain some magical hold over you, stem from the time when a person's name represented a characteristic. That characteristic defined the person under discussion, and if you knew that name you knew something, perhaps everything, about the person: "to name is to know."

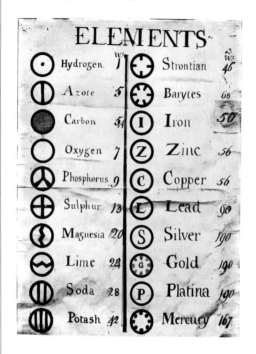

Figure 14.1: A table of chemicals; including some chemical element constructed by the father of modern chemistry, John Dalton (1766–1844); compare the symbols used for the entries in this table with the alchemical symbols, and planet symbols used by alchemists and astrologers (see Table 14.1, Figure 14.2, and Figure 3.1); for example, Hydrogen has the same symbol as the Sun (see Table 7.1), but Dalton could not have known that the Sun is composed mostly of Hydrogen. Image from: https://en.wikipedia.org/wiki/History_of_the_periodic_table.

Names are at the heart of any classification of the world. They are therefore at the heart of science. A true name is the name of an object, or an animal that expresses, or is somehow identical to the true nature of that object or animal. The notion that language, or some specific sacred language, refers to things by their true names has been central to a great deal of literature, philosophy, as well as various traditions of magic, religious invocation, and mysticism (mantras) since antiquity (see Chapter 2). And thus the idea that if you know a person's secret, or true name you can gain a magical power over that person, for example, the true name of the Egyptian Sun god, Ra, was revealed to Isis only through an elaborate trick. This knowledge gave Isis complete power over Ra, and allowed her to put her son Horus on the throne. Socrates in Plato's *Cratylus* considers, without taking a position, the possibility as to whether names are "conventional" or "natural"; that is, whether language is a system of arbitrary signs, or whether words have an intrinsic relation to the things they signify. Odysseus, when captured by Polyphemus in Homer's *Odyssey* is careful not to reveal his name; when asked for it, Odysseus tells the giant that he is "nobody." But later, having escaped after blinding Polyphemus and thinking himself beyond Polyphemus' power, Odysseus boastfully reveals his real name; an act of hubris that was to cause enormous problems later in the story. Knowing his name, Polyphemus was able to call down upon Odysseus the revenge of his father, the god of the sea Poseidon.

14.1   THE PROBLEM OF NAMING THINGS IN CONTEMPORARY SCIENCE

Today in astronomy, celestial objects are often named after individuals as we have exhausted the Classical pantheons of pagan gods and goddesses. It is likely that in some cases, using a personal name is justified. One can well imagine, for example, the relief of the European nations in September 1683 when the King of Poland, Jan (III) Sobieski, defeated the armies of the Ottoman Sultan, Mehmed IV at Kahlenberg, thereby saving Christian Europe from the Turks. In recompense, Sobieski's name was given to a newly discovered constellation of Stars, a rare honor for someone who did not live on Mount Olympus. However, when we learn that there are comet and small planet hunters who scan the heavens for new objects so that they may name them after family members and friends, one wonders at the seriousness with which they regard their enterprise. Astronomy is a field crying out for a totally rational system of nomenclature, although the International Astronomical Union does have a numbering system for asteroids, and has adopted a numerical system for comets, they continue to use a personal name in parentheses.

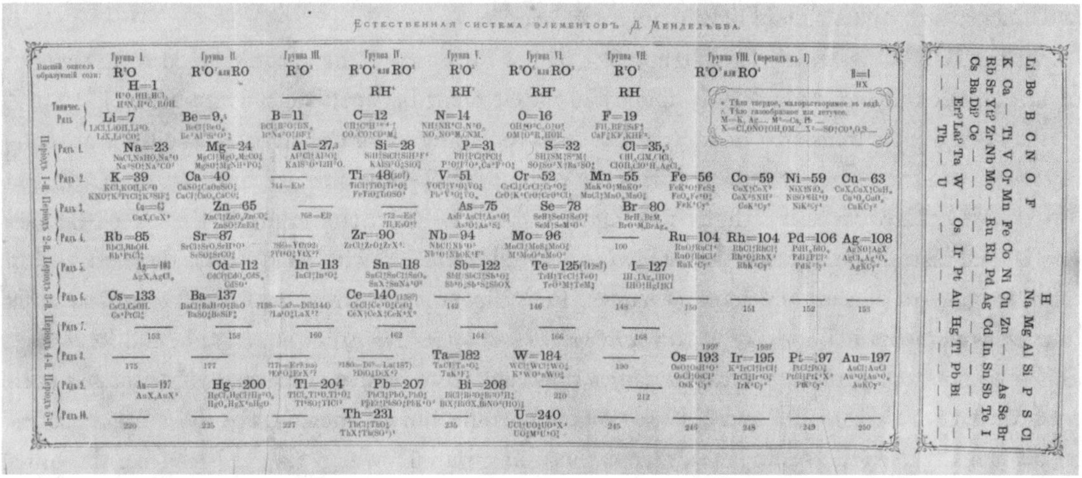

Figure 14.2: The first modern tabulation of the chemical elements, by Dimitri Mendeleev (1834–1907); dating from 1869–71. Image from: https://en.wikipedia.org/wiki/History_of_the_periodic_table.

Perhaps the most rarefied branch of experimental chemistry is the synthesis of new chemical elements. As one might imagine, this is not easy but nuclear chemists are continually attempting such syntheses. At present there are 118 named chemical elements, but the naming process is not always straightforward. And we will see that in an investigation of the relevance and appropriateness of naming new chemical elements after towns, laboratories, and scientists, one begins to question that oft-stated description of science as being supranational; particularly, science related to nuclear physics and the disintegration of radioactive nuclei.

The International Union of Pure and Applied Chemistry (IUPAC) is the body, which since 1919 has been charged with organizing a systematic nomenclature of chemistry, and this includes the naming of the chemical elements. Figure 14.2 gives a picture of the classification of the chemical elements in the middle of the 19th century. Today, the naming of a new element is complex. Gone are the days when early chemists such as Humphrey Davy (1778–1829) isolated whole columns of the Periodic Table,[38] and assigned the names we still use. Today, when a discovery is first published a temporary systematic element name is assigned, by IUPAC to the newly synthesized chemical element. In chemistry, a transuranic element (heavier than Uranium) receives a permanent name and symbol only after its synthesis has been confirmed by a second laboratory (there are only two or three laboratories in the world capable of doing these experiments). In some cases, however, such as the Transfermium Wars, controversies about priority and the naming of the elements have arisen; and there have been protracted international disagreements. Such controversies are not only deeply embarrassing for the laboratories concerned (questioning the science), but also embarrassing politically because non-scientists get involved to further their own ends by boosting national pride and chauvinism, which should have no place in science.

The IUPAC systematic, but temporary, names for a new element are derived from the element's atomic number, and are only applicable for elements between atomic number (Z) $101 \leq Z \leq 999$. Each digit is translated to a numerical root, according to published rules. The roots are concatenated, and the name is completed with the ending-suffix -*ium*. Some of the roots are Latin and others are Greek to avoid two digits starting with the same letter (for example, the Greek-derived *pent* is used instead of the Latin derived *quint* to avoid confusion with *quad* for 4). There are elision rules designed to prevent odd-looking names. The suffix -*ium* overrides traditional chemical suffix rules, thus elements 117 and 118 were *ununseptium* and *ununoctium*, not *ununseptine* and *ununocton*. This does not apply to the final trivial names these elements receive once their existence has been confirmed; thus element 117 and 118 are now Tennessine and Oganesson. For these trivial names, all elements receive the suffix -ium, except those in group 17 which receive -*ine* (like the other Halogens) and those in group 18 which receive -*on* (like the other Noble Gases). The systematic symbol is formed by taking the first letter of each root, converting the first to a capital. This results in three-letter symbols instead of the one- or two-letter symbols used for named elements.

---

[38] If you look at Mendeleev's Table of the Chemical Elements, Figure 14.2, you will see most of the elements isolated and named by Davy (sodium, potassium, calcium, strontium, magnesium, barium) in the first two columns or groups (Группа I and II).

After many years and a great deal of discussion in the appropriate IUPAC committee, the scientists responsible for the experiments that created these ephemeral species (few of the most recently discovered chemical elements exist for anything approaching a second; they are all radioactive species, and their half-lives are very short—usually fractions of a millisecond) will eventually agree on a true trivial name; that is, a name that can be used to identify this element in the non-specialist literature. Table 14.1 gives the names of the Transfermium chemical elements; Fermium is element number 100, and it can be clearly seen that these chemical elements have been named in accordance with the pre-eminent geo-political struggle of that period, the Cold War. There are about as many Russian as American names, with a few European names thrown in for good measure; to make it look as if it is not a Cold War club.

It is not for nothing that the two largest laboratories involved in this type of nuclear synthesis are the Lawrence Livermore (element 116) National Laboratory in California (element 96 is named after California) and the Flerov (element 114) Laboratory of Nuclear Reactions in Dubna (element 105) near Moscow (element 115), where the centers for Cold War research into nuclear weapons.

Table 14.1: Names and origin of the elements after Fermium in the Periodic Table of the Elements (see Figure 14.3)

Number in Periodic Table	Final Trivial Name (and symbol)	Origin of Trivial Name
100	Fermium (Fm)	Named in honor of the Italian-American physicist, Enrico Fermi (1901–1954).
101	Mendelevium (Md)	Named in honor of the Russian chemist, Demitri Mendeleev (1834–1907).
102	Nobelium (No)	Named for the founder of the Nobel prizes and armaments manufacturer, Alfred Nobel (1833–1896).
103	Larwrencium (Lr)	Named in honor of the American physicist, Ernest O. Lawrence (1901–1958).
104	Rutherfordium (Rf)	Named in honor of the British (New Zealand born) physicist, Lord Ernest Rutherford (1871–1937).
105	Dubnium (Db)	Named after the town of Dubna in Russia.
106	Seaborgium (Sg)	Named in honor of the American chemist, Glenn T. Seaborg (1912–1999).
107	Bohrium (Bh)	Named in honor of the Danish physicist Niels Bohr (1885–1962).
108	Hassium (Hs)	Named for the German state of Hesse

109	Meitnerium (Mt)	Named in honor of the German chemist, Lise Meitner (1878–1968).
110	Darmstadtium (Ds)	Named for the German city of Darmstadt.
111	Roentgenium (Rg)	Named in honor of the German physicist, Wilhelm Conrad Röntgen (1845–1923).
112	Copernicium (Cn)	Named in honor of the Polish astronomer, Nicolaus Copernicus (1473–1543).
113	Nihonium (Nh)	Named for Japan.
114	Flerovium (Fl)	Named for the Flerov Laboratory of Nuclear Reactions, Dubna, Russia.
115	Moscovium (Mc)	Named for the city of Moscow, Russia.
116	Livermorium (Lv)	Named for the Lawrence Livermore National Laboratory, CA, U.S.
117	Tennessine (Ts)	Named for the state of Tennessee, USA.
118	Oganesson (Og)	Named in honor of the Russian chemist, Yuri Oganessian (born 1933).

Speaking personally, the origins and meaning of the names given to the chemical elements is sometimes the first romantic attachment formed by young chemists with their future career. Many of us are fascinated by the way that the various names have been derived. For example, some are taken from the name of the mineral from which the element was extracted (for example, Sodium (from soda), Potassium (from potash), or Carbon; *carbo* being Latin for charcoal), and some from the locality where the mineral containing the element was found (for example, Strontium, from Strontian in Scotland, or Copper derived from *Cuprum*, the Latin name for Cyprus,). Other names of elements derive from the name of the city where the discoverer lived (for example, Hafnium, *Hafnia* being Latin for Copenhagen), or from the color of the purified element (for example, Chlorine, from *chloros*, meaning greenish-yellow in Greek).

The element's name might also be derive from some characteristic of the chemical properties of the element, thereby providing information about its chemistry. This may include inertness (for example, Argon; in Greek *argos* means inactive), or reactivity (for example, Bromine, *bromos* being Greek for stench, or Fluorine, where in Latin *fluere* means flux). Indeed, the name may even derive from the difficulty of extracting the element from the naturally occurring source, again imparting important chemical information. Examples of these include the Greek *lanthanein* (Lanthanum, element number 57) which means to "lie hidden," and *dysprositos* (Dysprosium, element number 66) meaning "hard to get at." If one is familiar with the Greek myths, it is easy to understand why Niobium (element number 41) and Tantalum (element number 73) are so named. Niobe was the daughter of Tantalus, and the two elements are found together in the same ore. It was only in 1844 that they were shown to be two distinct elements. The element Tantalum was first isolated in 1802, but Niobium was not isolated until 1864, when it was extracted from a purified ore of Tantalum.

Sir Humphry Davy and Jöns Jacob Berzelius (1779–1848) isolated and named almost whole columns of chemical elements. Davy was the first to isolate and name eight of the elements: Boron, Barium, Calcium, Chlorine, Magnesium, Potassium, Sodium, and Strontium, whereas Berzelius only managed to isolate and name four elements: Cerium, Selenium, Silicon, and Thorium. Other well-known chemists who figure in this list of discoveries include Friedrich Wöhler (1800–1882), who isolated Aluminium and Beryllium, and Robert Wilhelm Bunsen (1811–1899) who, in a triumph of early analytical chemistry, spectroscopically identified Caesium and Rubidium, without even isolating weighable quantities of the pure metals. Bunsen named these elements from the color of the principal spectral line; *caesius*, Latin for sky-blue, and *rubidus*, Latin for deepest-red. These early chemists did not simply add extra elements to the list of elements already known, they put detailed information about the structure and properties of the newly discovered elements into the new names; just as Linnaeus did in his system of binomial nomenclature of organisms (see Chapter 13).

Recently, however, there has been a trend to name elements after individuals. Cynics might say that this trend has arisen because few of today's nuclear chemists know the Greek myths, or any Classical languages. Probably the last element named from a distinctive characteristic, as opposed to merely adopting the Latinized name of the university or state where it was discovered, was the man-made metal Technetium (element number 43), discovered in 1937 and named, appropriately, from the Greek *technetos*, meaning artificial; although Dimitri Mendeleev knew that such an element must exist when he was putting together the first pictorial representation of the Periodic Table, he left a gap for it (see Figure 14.2 where the gap is at number 44).

Unfortunately, the modern desire to name elements after scientists has allowed politics and nationalism to creep into the Periodic Table. During the Cold War, Russian scientists suggested the name of an eminent Russian scientist for an element they claimed to have discovered, U.S. scientists suggested the name of a U.S. scientist, and Germans suggested a German scientist, or town. As a result, the naming of the most recently discovered chemical elements required more international compromise than classical erudition. Such national disagreements over the name of a chemical element are unfortunately not new, and can never improve the public perception of science or, more importantly of scientists. In 1950, IUPAC had to intervene, after almost a century of controversy, to recommend that the name of element number 41 be Niobium. The first specimen of the ore containing this element was found in the American Colonies by John Winthrop (1714–1779), professor of natural philosophy at Harvard College, but this specimen was sent to England for study. However, many American institutions continued, after its isolation in 1864, to refer to this element as Columbium (Cb), after the spirit of America. Nevertheless, it is the name Niobium that is today universally accepted by working scientists.

14.2   THE TRANSFERMIUM WAR

Given the nature of American-Soviet rivalry in the forty years following the World War II, it is perhaps not surprising that this rivalry carried over into the world of scientific research. Although given that the particular area of research, which interests us here involved the investigation of the stability of atomic nuclei, and that the research was largely carried out in laboratories better known for their research into developing nuclear weapons, this Cold War rivalry is probably not unexpected. But there was more to this competition than simple, international political posturing; there also no small measure of personal vanity.

After all, the naming of new chemical elements is a rare event, unlike the discovery of a new asteroid or a new comet. And if your name is chosen, you join one of the most hallowed clubs in all of science. The lucky few who are immortalised by having their name adopted as a chemical element. Consequently, the names for the chemical elements beyond number 100 were the subject of a major international controversy starting in the 1960s, described by some nuclear chemists as the Transfermium War. This controversy was only resolved in 1997, and only by a significant weakening of one of the international organizations created early in the 20th century to permit science to continue internationally, even when nations were at war.

The controversy arose from disputes between American scientists and Soviet scientists as to which had first made these particular elements. One cannot be said to have isolated these elements as Humphrey Davy did back in the late 18th century as they are ephemeral; they are unstable and their half-life is usually only a few milliseconds, or even a few microseconds. Consider element 112, Copernicium ($_{112}$Cn). This element was first observed in the Heavy Ion Research Laboratory (*Gesellschaft für Schwerionenforschung*, GSI) in Darmstadt (element number 110), Germany (element number 32), where researchers were attempting to fuse the nuclei of Lead and Zinc (named by the alchemist Paracelsus after the form of its crystals) atoms. The reaction scheme for the generation of element 112 and its subsequent decay via neutron and alpha-particle emission are given in the following scheme (where the superscripted number refers to the atomic weight of a particular isotope of the element, and the subscripted number refers to the element's atomic number): $^{208}_{82}$Pb + $^{70}_{30}$Zn → $^{278}_{112}$Cn → -($^{1}_{0}$n) [loses a neutron] → $^{277}_{112}$Cn → -(alpha) [losses an alpha-particle[39]] → $^{273}_{110}$Ds → -(alpha) [losses another alpha-particle] → $^{269}_{108}$Hs → -(alpha) [losses another alpha-particle] → $^{265}_{106}$Sg → -(3 alpha) [losses three alpha-particles] → $^{253}_{100}$Fm.

The scientists undertaking these experiments observed a single atom of $^{277}_{112}$Cn on February 9, 1996. It should be noted that $^{277}_{112}$Cn decays after 280 microseconds, $^{273}_{110}$Ds decays after 110 microseconds, and $^{269}_{108}$Hs is relatively long-lived, decaying after 19.7 seconds. These are all ephemeral materials made in quantities so small they cannot be weighed. The same research group

---

[39] An α-particle (alpha-particle) is a doubly-ionized Helium atom; that is, He^{2+} (a bare Helium nucleus) with a mass of 4 amu (or 6.644 657 230(82) × 10^{-27} kg)

were the first to observe element 111 by fusing Bismuth nuclei with Nickel nuclei, and were rewarded by seeing three atoms of the desired product over the period December 8–18, 1994.

By convention, the right to suggest a name for a newly discovered chemical element goes to their discoverers. However, for the elements 104, 105, and 106 there was a controversy between a Soviet/Russian laboratory and an American laboratory regarding priority. Both parties suggested their own names for elements 104 and 105, neither recognizing the names suggested by the other laboratory. This is what brought IUPAC into the debate. In addition, the American name of Seaborgium for element 106, chosen by the American Chemical Society to honor Glenn T. Seaborg (1912–1999) of University of California, Berkley (a Nobel laureate chemist who had also been a science adviser to U.S. presidents during the Cold War) was objectionable to some, because it referred to an individual who was still alive at the time his name was proposed. Einsteinium (element number 99) and Fermium (element number 100) had also been proposed as names for new elements while Einstein and Fermi were still living, but by the time that the names of these two eminent physicists were adopted, both scientists were dead. So, there was no precedent at this time for naming a chemical element after a living person. However, Seaborg wanted this fame while still living. And this caused serious international tensions; reviving much of the rivalry that had existed during the Cold War, and with no little Cold War rhetoric.[40] In addition, the Soviet Union wished to name element 104 after Igor Kurchatov (1903–1960), builder of the Soviet atomic bomb, which was another reason the name was objectionable to Americans.

The two principal groups which were involved in the conflict over element naming were: an American group at Lawrence Berkeley Laboratory, California, and a Russian group at Joint Institute for Nuclear Research in Dubna, Russia. And between these two national teams, the referee, or arbiter, was the IUPAC Commission on Nomenclature of Inorganic Chemistry, which introduced its own proposal to the IUPAC General Assembly (the Union's highest decision making body) for the names of these elements. The German group at the GSI in Darmstadt, who had undisputedly discovered elements 107 to 109, were dragged into the controversy when the IUPAC Commission suggested that the name "Hahnium" (in honor of the German physicist Otto Hahn (1879–1968), who won the Nobel physics-prize in 1944 and, although opposed to the Nazi Party, had remained in Germany throughout WWII); a name already proposed for element 105 by the Americans, be used instead for GSI's element 108. In short, no national laboratory was happy, and it was in fact a major blow to the prestige of IUPAC.

In 1994, the IUPAC Commission on Nomenclature of Inorganic Chemistry proposed the names given in column six of Table 14.2. Thus attempting to resolve the international disagreement by sharing the naming of the disputed elements between Russians and Americans, replacing the name for 104 with one honoring the Dubna research center, but not naming 106 after Seaborg.

---

[40] The author was, at this time, the Deputy Executive Secretary and editor of IUPAC and saw, read, and heard the voluminous correspondence and exchanges concerning this sad affair.

However, this solution drew objections from the American Chemical Society on the grounds that the right of the American group to propose the name for element 106 was not in question, and that group should have the right to name the element. IUPAC further confused things by deciding that the credit for the discovery of element 106 should be shared between Berkeley and Dubna, but the Dubna group had not come forward with a name. Along the same lines, the German group protested against naming element 108 with the American suggestion Hahnium, mentioning the long-standing convention that an element is named by its discoverers. In addition, given that many American textbooks had already used the names Rutherfordium and Hahnium for elements 104 and 105, the ACS objected to those names being used for other elements.

Finally in 1997, the 39th IUPAC General Assembly in Geneva put forward the names given in the last column of Table 14.2. Professor Glenn Seaborg died in 1999, however this attempt at creating a tradition of naming chemical elements after living people has continued with the Russian chemist, Yuri Oganessian whose name is given to element 118, Oganesson. Thus, the convention of the discoverer's right to name their elements was respected for elements 106 to 109, and the two disputed claims were shared between the two opponents.

Table 14.2: A summary of the evolution of the names of some of the transfermium elements

Atomic Number	Systematic IUPAC Name	Proposed American Name	Proposed Soviet/ Russian Name	Proposed German Name	Suggested IUPAC in 1994	Final Name (IUPAC 1997)
104	unnilqua-dium	Rutherfordium	Kurchatovium	-	Dubnium	Rutherfordium
105	unnilpentium	Hahnium	Neilsbohrium	-	Joliotium	Dubnium
106	unnilhexium	Seaborgium	-	-	Rutherfordium	Seaborgium
107	unnilseptium	-	-	Neilsbohrium	Bohrium	Bohrium
108	unniloctium	-	-	Hassium	Hahnium	Hassium
109	unnilennium	-	-	Meitnerium	Meitnerium	Meitnerium

This modern personality cult is inappropriate and inherently nationalistic, laying itself open to political problems. It was a lot simpler, and more appropriate, when the names of mythological characters or names derived from chemical properties were used for the elements. Myths and legends are the common heritage of all mankind and tell us, by analogy, more about the element, for example the chemical affinity between Niobium and Tantalum, than do Fermium or Nobelium, which were never associated with Enrico Fermi or Alfred Nobel. And unlike Niobium—a relatively common, naturally occurring, element whose salts are key materials used in modern electronics—element 106 has a half-life of a few hundred microseconds and will only ever be available in the minutest of quantities.

The name Iridium, derived from the Latin iris, meaning color, as exemplified by the salts of this element, and Iodine, from the Greek iodes meaning violet; both impart chemical information. Likewise with Antimony, derived from the Greek *anti monos*—"a metal not found alone," *savants* of the Ancient World are telling us that this element is unreactive enough to be found as a native metal, but always associated with its chemically similar neighbors in the Periodic Table. This is quite a lot of information for the Ancient World (Antimony salts were used as cosmetics by the Ancient Egyptians). In comparison, the names Tennessine, Nihonium, Hahnium, and Meitnerium tell us nothing and create confusion because we would need to consult textbooks of history, or English–Japanese dictionary to identify the origins of their names—let alone their discoverers. Berzelius refused to name elements after people; when the discoverer of Tungsten (element number 74), Carl Wilhelm Scheele, was to be immortalized as the discoverer of this new element, Berzelius remarked "*The immortality of our compatriot does not need this support.*" Thus, today we have Tungsten and not Scheelium.

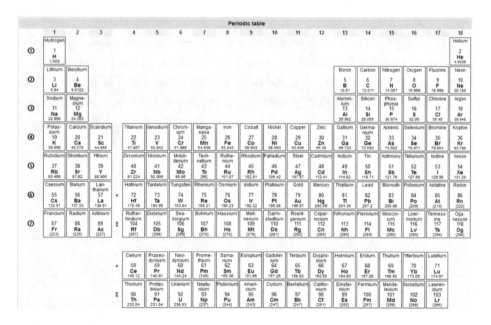

Figure 14.3: The well-known Periodic Table of the Elements (source: https://en.wikipedia.org/wiki/Periodic_table). Hydrogen (H) is element number 1; Uranium (U) is number 92; Iron (Fe) is in the middle at number 26. Compare with the image in Figure 14.1; it took two centuries for scientists to shake off the last vestiges of alchemy.

However, many modern scientists prefer the names of scientists as labels for the chemical elements. Unfortunately, the choice of scientist to be so honored is arbitrary and illogical—why choose Lawrence, who was not a chemist, while Humphry Davy, who discovered eight elements, and Frederick Soddy, who discovered and explained the existence of isotopes, have not been so honored? If only truly great scientists are to be so honored, why not Newton, Maxwell, Faraday, or Galileo?

14.3   FURTHER READING

The International Union of Pure and Applied Chemistry (IUPAC) was founded in 1919 by chemists from industry and academia who recognized the need for international standardization in their area. As I have pointed out in this volume, the standardization of weights, quantities, names, and symbols is essential to the successful advance of the scientific enterprise, and to the smooth development and growth of international trade and commerce.

IUPAC is the authority on chemical nomenclature and terminology, and two IUPAC bodies take leading roles in this activity: Division VIII—Chemical Nomenclature and Structure Representation and the Inter-divisional Committee on Terminology, Nomenclature, and Symbols (see https://iupac.org/ for more details). As one of its major activities, IUPAC develops Recommendations to establish unambiguous, uniform, and consistent nomenclature and terminology for specific scientific fields, usually presented as: glossaries of terms for specific chemical disciplines; definitions of terms relating to a group of properties; nomenclature of chemical compounds and their classes; terminology, symbols, and units in a specific field; classifications and uses of terms in a specific field; and conventions and standards of practice for presenting data in a specific field. Information on chemical terminology can also be accessed through the IUPAC Color Books, which may be consulted on-line at https://iupac.org/what-we-do/books/color-books/.

CHAPTER   15

The Evolving Science of History

Social media: Websites and applications that enable users to create and share content or to participate in social networking.

The premise of this volume is that scientists (but not social scientists) have a worldview that is different from that of non-scientists, particularly, for example, historians. Whether this is a good thing or a bad thing, given the predictive power of science, it is irrelevant. Science is in the world, and it cannot be removed. We have seen how science has evolved over the last few millennia, how it has extended human life, and how it is capable of extending it a lot further. There is no problem that has arisen from some aspect of the misuse of science that cannot be corrected by application of more science. This is true whether we are considering nuclear power, or the influence of man-made greenhouse gases in the Earth's atmosphere. Our errors catalyze future progress.

Ordinary people may not understand science, particularly, the physical sciences, and they may also be in awe and fearful of the power of science and of the scientist, but it is to science and the scientist that politicians must turn when they need a solution to a technical (non-social or non-political) problem. After all, there is no one else to turn to; and science has already transformed the world and society on a number of occasions. For example, the rise of science and medicine in the early-modern age of the 17th century; the Victorian Internet of the international telegraph (William Thomson became Lord Kelvin for his invention of the equipment needed to lay conducting cables between the UK and the U.S.); atomic power; genetic medicine; and the creation of the modern Internet.[41] These things cannot be undone; scientific discoveries once made cannot be forgotten. Society will have been transformed, and if the technical details of an advance become lost; then it will survive in the form of legend and myth, which will catalyse its re-invention. History waits for no one; it is always on the move, somewhere. The Second Law of Thermodynamics tells us that the arrow of time can only point in one direction—into the future. Scientific advances are the ratchets in the mechanism of history that prevent history, and social advance running backwards.

But what of history, that subject that fascinates all thoughtful individuals? Anyone who has read, at least, two history texts by different authors on the same period of history will know that historians can be maddening people. Yet asked if history has a pattern or a plan, they will usually assert that such a question is offensive; history is but the record of the chaotic events that come about because people's hopes and ambitions are invariably modified by external circumstances—

---

[41] Question: how did we organize holidays and trips to the theater before the invention of the Internet?

often the hopes and ambitions of other people. As Edward Gibbon put it, *"History is but the record of crimes and misfortunes."*

You will likely be told that an historian's job, is not to find a pattern, let alone a fundamental law, but to ensure by research that the record is accessible and intelligible. There are, however, various schools of historians. There are Marxist historians, who look preferentially for patterns in the data of events, which they believe reveal the signature of the perpetual struggle between the proletariat and their economic masters. Other historians hold to the idea of progress, or the notion that the improvement of the human condition in recent centuries can, with ingenuity be extrapolated into the future. Physical scientists who have an interest in history often fall into this camp. Without realizing it, however, these physical scientists are adopting the ideas of the science fiction writer, Isaac Asimov (1920–1992); best expressed in his invented science of psychohistory. That the future may be predicted, if only we had sufficient data and powerful enough computers.

Irrespective of the desires of even the most ardent of anti-scientists, whether their dislike of science arises from their religion, their politics, or their limited education, science is here to stay. As an example of the now unavoidable influence of the mathematical or physical sciences on society, let us consider the mining of personal data on social media. This study will also demonstrate how techniques of mathematical physics are used to analyze a set of data. This is a topic that will likely remake our society over the next generation. It will certainly do away with conventional politics; there will still be elections in the future, but there will be no need to go to a polling office to vote. We will all like or dislike a particular politician, or a particular proposal for a law on social media.

15.1   SOCIAL MEDIA

While it may seem premature to associate history with something as new and vibrant as Facebook, it must be commented that a great many people, of a variety of political views, religious persuasions and commercial interests are sifting through; that is, mining the data we have all (well over 2.5 billion of us… and increasing) carelessly left littering social media.

Before writing was invented, there was only one way for an individual to leave behind a record of himself for future generations. They would place their hand on cave walls and blown a mouthful of pigment over it; leaving behind a stenciled handprint. It was a successful strategy, as a great many of these haunting cave paintings survive. With time, our civilization evolved, as did an individual's ambitions and desires. We have now become amazingly adept at recording our lives. We have built mausoleums and libraries, and filled those libraries with books; we have written books of history and compiled sophisticated records of our ancestors. Then a college student developed an extraordinarily simple and useful tool to convey our personal histories, and interests to future generations—Facebook.

Historians have always struggled to tell the stories of our everyday ancestors, even those who lived only a few generations ago. Historical records offer great insight into a handful of important and powerful people, but piecing together the lives of ordinary people has always been difficult. Facebook changes all this. In little more than a decade, Facebook's users have contributed to a massive depository of personal information that documents both our reactions to events and our evolving customs with a scale and intimacy earlier historians could only dream about. It's hard to estimate just how substantial this database of personalities could become. Presently, more than 2.5 billion people are regular users of Facebook. Assuming people will continue to use the site regularly, this means most of these users will document more of their lives over the coming years, leaving behind photos, details of friendships and love affairs, their likes and dislikes, and their reactions to news-worthy events. In addition, there are tens of millions of deceased Facebook users; individuals who have left behind digital remains. It is estimated that if Facebook stopped growing tomorrow, the number of deceased users "on" (perhaps one should say "in") the site would be well over a billion by the end of the century. If the site were to continue growing as it is now, that number could reach about 5 billion users by 2100. All this data, this information is there to be mined and exploited—for whatever reason.

Today, these "dead" accounts offer a virtual environment for mourning. However, the data they contain will be invaluable to future historians and sociologists. They will be able to investigate the events surrounding the election of Donald Trump, and the online culture wars that facilitated and followed this election. Similarly, with the great saga of Brexit, future historians will be able to study what ordinary people thought about Brexit and the politicians who desired Brexit, rather than merely what the politicians who desired Brexit thought about Brexit. Of course, future researchers will see lots of pictures of dead cats and puppies, of homemade cakes, and *Game of Thrones* memes with which users distracted themselves from the daily grind of work. But, this potential utility complicates, rather than resolves, the problems of security already plaguing Facebook: How much privacy do the dead deserve? How do we guarantee that those who wish to be forgotten are allowed this right—is it a right? This matters because, for the first time, we all have the power to leave behind far more personalized histories than any previous generation. We don't have to rely on the recollection of our descendants for our memory to survive, and we don't have to accept that our collective experiences will fade away with time. This is our chance to leave far more than our handprints on the digital walls of history.

Let us now look briefly at the possible misuse of the data of the living, and of the dead. What is happening to our data, to our likes and dislikes, our loves and hates that we all leave all over social media? If I, for example, like a friend's post about what he and his boyfriend did on the weekend, who outside of a list of our mutual friends looks at this information, and for what purpose may this information (data) be used by parties unknown to anyone actually involved? If these data miners restricted themselves to collecting recipes for chocolate cookies, there would be no problem, but it appears that this is not what data mining on social media is all about.

15.2   SOME DETAILS OF THE ANALYSIS OF PERSONAL DATA ON SOCIAL MEDIA

The first study of Facebook data, that is, data mined from the personal profiles of Facebook users, was based on a sample of 58,466 volunteers (thus in Table 15.1, $N = 58,466$) from the U.S.; this data was obtained through the myPersonality Facebook application (no longer available), which included their Facebook profile information, a list of their likes (with, on average $n = 170$ likes per person; totalling about 10 million likes, and one assumes dislikes), psychometric test scores, and survey information. Users and their likes were represented as a sparse user–like matrix, the entries of which were set to 1 if there existed an association between a user and a like, and to 0 otherwise.

The modelers selected a wide range of traits and personal attributes from users that clearly reveal how accurate and potentially intrusive such an analysis could be; these characteristics included: sexual orientation, ethnic origin, political views, religion, personality, intelligence, satisfaction with life, substance use (alcohol, drugs, cigarettes), whether an individual's parents stayed together until the individual was 21, and basic demographic attributes such as age, gender, relationship status, and size and density of the friendship network. This data was then represented in the form of a matrix (see Table 15.1), and standard statistical methods of analysis were used for their manipulation.

Table 15.1: A representation of the raw data mined from social media sites such as Facebook. The N users are represented as the first column, and the various characteristics of each of these N users form the columns of the table or matrix; for example, the user either likes (1) Mickey Mouse or he/she dislikes (0) Donald Trump. There can be as many columns as there are data-fields available for personal traits and characteristics: ethnicity, sexual orientation, religion, etc. [1]

	Likes Art	Likes Gay-sex	Likes Christianity	Likes Evangelical Christianity	Likes Donald Trump	Likes Brexit	Admits to voting for Brexit	Likes drugs	Likes demonstrating	Likes Mickey Mouse
Facebook user 1	0	1	0	0	1	0	1	1	1	1
Facebook user 2	0	1	1	0	0	0	1	0	0	1
....										
Facebook user N	1	0	1	0	0	1	0	1	0	1

Matrix decomposition, also known as matrix factorization, is the standard initial procedure in the analysis of any large body of data from which researchers wish to make predictions. Perhaps the best-known and widely used matrix decomposition method is the singular-value decomposition (SVD). All matrices have an SVD, and as such it is used in a huge range of applications. The SVD takes a rectangular matrix of, for example, likes and dislikes from a selection of users of Facebook; defined as **M**, where **M** is an *m* by *n* matrix, see Figure 15.1, in which the *m* rows represents the *N* social media users being studied, and the *n* columns represents those users' likes and dislikes (see Table 15.1). The SVD theorem states:

$$\mathbf{M} = \mathbf{U}\,\Sigma\,\mathbf{V}^{*}$$

Here, **U** is an *m* by *n* unitary matrix; Σ is a diagonal *m* by *n* matrix; **V** is a symmetric *n* by *n* matrix, and **V*** is the conjugate transpose of **V**. The SVD represents an expansion of the original data in a coordinate system where the covariance matrix is diagonal, and so more amenable to statistical analysis.

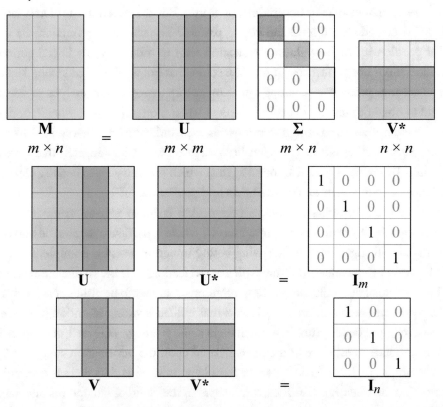

Figure 15.1: Details of the matrix algebra involved in the singular-value decomposition by which the raw data of our likes and dislikes on social media are turned into a set of linear equations, from which the likelihood of our "behavior" may be calculated. (The lowercase, italicized letters denote the type of matrix—symmetric or non-symmetric.)

Having now prepared the data in a standard statistical format, it is possible to make predictions about the group under study; the users of social media such as Facebook, as a representation of humanity in general, or just the voting population of the U.S. or the UK. In statistical modeling, regression analysis is a set of statistical processes used for calculating the relationships among variables. It includes many techniques for modeling and analyzing several variables, when the focus is on the relationship between a dependent variable and one or more independent variables (or predictors). More specifically, regression analysis helps one understand how the typical value of the dependent variable changes when any one of the independent variables is varied, while the other independent variables are held constant.

Let us consider an example of such an analysis that could be readily applied to data available on social media: *the influence of education upon voting intention*. To obtain the raw data we take a large sample of individuals of similar age and ask them how long they spent in full-time education and what are their political likes and dislikes. Clearly, when graphed such raw data will generate a plot with an enormous range of uncertainty, or scatter. But the question of interest is: Is there a relationship between education level and voting pattern? The scatter of points on the graph may suggest that people with higher values of education tend to follow a more liberal voting pattern, but the relationship is not perfect; and it would be clear that knowledge of education level does not suffice for an accurate prediction of voting intention. To apply regression analysis to this particular problem, and to obtain a smoother curve of the data when plotted (for better predictions) requires that we first hypothesize that voting intentions for each individual are determined by education, and by a collection of other factors (race, locality, sex, sexuality, profession, etc.) that we term "contributing noise." This noise is the background against which one is trying to derive a clear cause and effect relationship, upon which to base extension and prediction.

First, we write a hypothesized model relationship (a linear regression model) or equation between education level (E) and voting intention (I, where a particular numerical value, or range can be correlated with right-wing or left-wing voting intentions) as, for example, $I = \alpha + \beta E + \varepsilon$, where α is a constant (how one would vote with zero education), β is the coefficient relating how an additional year of education influences voting intention (assumed to be either positive or negative), and ε is the noise term representing other factors that influence voting intention (for example, age, postal address, religious beliefs, etc.). The parameters α and β are not observed, and regression analysis is used to produce an estimate of their value, based upon the information contained in the raw data set; for example, in Table 15.1. What we have hypothesized is that there is a straight line, or linear relationship, between E and I. Thus, somewhere in the cloud of data points derived from our social media user-like matrix we expect to find a line defined by the equation $I = \alpha + \beta E + \varepsilon$. The task of estimating α and β is equivalent to the task of estimating where this line is located relative to the axes, I and E. The answer depends in part upon what we think about the nature of the noise term, ε. To estimate ε, one would look at the solutions (eigenvalues) of the matrix of likes and dislikes for various groups of individuals. The predictive modeling undertaken by organizations such as Cambridge Analytica (now bankrupt) involved fitting various models of correlations (such as education, age, or sexuality with voting intention, or liberal sentiments, or lack of liberal sentiments) using standard techniques of linear regression. Variables such as age or intelligence were predicted using a linear regression model, whereas dichotomous variables such as gender or sexual orientation were predicted using logistic regression [2].

So how close are we to a precise science of prediction based on the analysis of the personalities of individuals comprising large groups? To my mind, the present furore about the actions of the now defunct company Cambridge Analytica in the UK and the U.S., and Facebook in seeking to gain personal data on hundreds of millions of individual voters with the desire to target individual voters with the most appropriate publicity material, so as to influence elections, and thereby direct the course of history, is merely a first overt attempt at creating something akin to Isaac Asimov's psychohistory. But perhaps it is too premature—it only just worked and has been revealed to public scrutiny and condemnation, which was probably not the intention. Perhaps there are not yet enough individuals on social media; we are still a long way from the assumptions of Asimov about the appropriate size of the group being examined and modeled. But this new form of democracy is only in its infancy; it works, and that is a great stimulant for further development. Intrusion into one's Internet-space, is happening everywhere, and will only become worse.

Sadly or happily, depending upon one's point of view, we have a long way to go before we can approach the perfect state required for a true statistical modeling of society. The transition from the non-statistical behaviour of individual molecules (or individual humans in Asimov's fiction) to the more mathematically friendly statistical behaviour of large groups of molecules, i.e., solids and liquids (or human society in Asimov's fiction), is not so easily identified. However, the American historian Henry Adams (1838–1918), the grandson and great-grandson of American presidents, attempted such a modeling of human history at the beginning of the last century. In his *Degradation of the Democratic Dogma* (1919), Adams proposed two laws of history: *All civilization is centralization* and *All centralization is economy*. It is difficult to find fault with the first law; however, the second law says that resources; particularly, energy sources must be adequate to sustain the energy needs of the civilization or empire. Therefore, all civilization is the survival of the most economic system; for example, the nation that has an ample source of energy (coal, oil, gas, nuclear, etc.) and is able to control access to all major sources of energy for all other nations will necessarily dominate the world. There is a strange closeness between physics and history; a closeness that always moves out of focus when you seek to examine it in detail. In both physics and history, all is cause and effect; in history as in physics, there is no action without a reaction. The problem is that in any predictive, quantitative estimation derived from history and from physics, the error bars are larger for the former than for the latter.

15.3 FURTHER READING

1   The full details of such analyses are to be found in Private traits and attributes are predictable from digital records of human behavior; Michal Kosinskia, David Stillwella, and Thore Graepel; *Proceedings National Academy of Sciences* (2013); 110(15): 5802–5805.

2   This topic is discussed in detail in any textbook on the statistical analysis of experimental data; for example, *Linear Algebra and Matrix Analysis for Statistics* (2014): Sudipto Banerjee and Anindya Roy; Texts in Statistical Science; Boca Raton, FL:Chapman and Hall/CRC Press. And *Quantifying Measurements: The Tyranny of Numbers*; Jeffrey H. Williams (2016); San Rafael, CA:Morgan & Claypool, and references therein.

CHAPTER 16

Obfuscation: Why Are We Not Living in a New Golden Age?

Where is the wisdom we have lost in knowledge, and the knowledge we have lost in information?

T.S. Eliot, *What is a Classic?*;
Presidential address to the Virgil Society of London, 1944.

In this book, I have tried to show the evolution of science was an attempt to classify, understand, and contain Nature. Originally, this involved the creation and memorising of long lists of natural events, things and phenomena; then we abandoned the lists and looked for the coherences that existed behind the observations. We searched for the rule or universal law that gave rise to, or led to, the observed thing or phenomenon. Why man wished to understand his environment is simply stated; he wished to protect himself and his extended family, or tribe in a hostile and indifferent world. We invented religions and social systems to further enable this protection. Science, as we know it today, grew out of the failure of religion and magic, that is, pre-scientific natural philosophy, to explain and predict natural events. Science worked on a regular basis, unlike magic and religion that only worked on a statistical basis. But this triumph of science has not eliminated magic and religion; they are still with us today, and in the case of religion still play an important role in the stability of the wider society.

All this effort to create a scientific worldview was directed toward improving the lot of humanity. Men imitated the divine lawgivers in our mythologies and religions, demonstrating that religion always develops before science in an evolving society, and thereby formulated the idea of Laws of Nature, which we all had to obey. Scientists believed that a codification and explanation of natural laws would help man return to that Golden Age when everything was perfect; when men were at peace with each other, and at one with Nature. This search for a better world goes on. The ultimate discovery, the Theory of Everything should enable man to cease struggling against Nature, as all would finally be revealed; there would be no questions left unanswered. Figure 11.2 shows that we are well along on the roadmap to the Theory of Everything. So, the question we must now ask is why are we not living in a new Golden Age? What has gone wrong? Or, are we living in a

Golden Age, and we just haven't noticed?[42] What happened to all that optimism and enthusiasm that gave birth to the first tentative steps of the great endeavor of science?

16.1   THE SCIENCE WARS

Sadly, it is true to say that not everyone wishes to be liberated from their narrow, limited, obscure way of looking at Nature. It is not universally assumed that scientists have the only correct way of interpreting the world we see around us. The fractious disputes or disagreements between scientists and some non-scientists as to who has a monopoly on objective truth have always generated a lot of heat. Scientists, particularly physical scientists, find sociologists, historians, and "critics of culture" tedious in their attempts to rub the gloss off the enterprise of science. In good relativistic fashion, non-scientists usually look askance at scientists' claims of an absolute objectivity and the ultimate truth, and regard scientists in the same way they would regard any other anthropological group. Social scientists often regard scientists as a tribe, whose structures of supposed authority, of peer review, as a means of maintaining their integrity, of training and initiation and peer-acceptance, should be subjected to the same levels of criticism, as would be the behaviour of any tribe of non-industrialised people to be found in the Amazonian jungle. Needless to say, this does not go down too well with the professional tribe who have given us: atomic power, aeroplanes, antibiotics, weapons of mass destruction, mobile phone, spaceflight, and genetic engineering. But then to contemporary cultural critics, such as the French philosopher and literary theorist Jean-Francois Lyotard, science is nothing more than another grand narrative, with a structure just like, and no better than history, sociology or semiotics. That is, science should be treated as an object of study in itself, not as an enterprise that provides a transcendental view of Nature.

Scientists (especially physicists) will, however, tell you that their view of Nature is the real, or the most real view of the world around us, that it is possible to gain. Scientists like to call themselves realists, but even some of the greatest of physicists have had doubts as to the transcendental viewpoint of the scientist. Niels Bohr, one of the founders of quantum mechanics, said that "*It is wrong to think that the task of physics is to find out how nature is. Physics concerns what we can say about nature.*" Nothing transcendental here; description, rather than explanation. This is a more realistic viewpoint; but, unfortunately, not one shared by all scientists.

So, when it comes to understanding the world in all its wonderful complexity, should we believe our ardent, zealous physical scientists, or the majority of mankind? The philosopher, physicist Karl Raimund Popper (1902–1994) held that all science provides are hypotheses that have, so far defied attempts to falsify them. Of course, this argument does not take account of the "trust" we put in the technological products of science. We design things not in accord with a fundamental

---

[42] Granted we have yet to find the Theory of Everything, but we are well on the way to glimpsing its spectral form in extreme events in Space and here on Earth, as in the discovery of gravity waves and of the Higgs boson.

principle of a theory, but with a theory that has survived rigorous examination, and which we fully expect to continue to survive ever more detailed examination. When you agree to have major surgery, you expect to wake up cured, and when you buy an airline ticket to Australia, you expect to get there, and so you buy a return ticket.

Modern medicine is successful, because it is based upon the scientific method of observation, hypothesis and testing. When we design jet engines, we do so in the context for which they are intended. In saying these things, we are not able to step outside our skins, or outside the Universe and attempt a majestic, dispassionate, transcendental examination. We are merely repeating science's own explanation of events and observations. There is a world of difference between observing and recording something, and an *a priori* explanation of a phenomenon. There is, in fact, no getting behind the explanation of science, because we are in the world we describe, we are part of it and cannot stand outside that world. As modern quantum mechanics tells us, the experimenter is part of the experiment; Schrödinger's cat again. Whereas a sociologist does often stand outside of society and deliver himself/herself of sweeping generalisations.

16.2   ANTI-SCIENCE

The problem between scientists and non-scientists is, today, often termed anti-science; this is a term that has arisen in our contemporary world of alt-truth and climate change sceptics. Anti-science is an extreme form of a suspicion of science; a position that rejects science and the scientific method. People holding antiscientific views do not accept science as an objective method that can generate anything of use to them—let alone universal knowledge. The more thoughtful, contend that scientific reductionism is an inherently limited means to reach an in-depth understanding of a complex world in continuous evolution.

At the beginnings of the scientific revolution, proto-scientists or *savants* such as Robert Boyle found themselves in conflict with non-practical thinkers, such as Thomas Hobbes, who were skeptical as to whether or not science was a satisfactory way of arriving at real, or genuine knowledge of the world. Hobbes' stance is sometimes regarded as an early anti-science position. And we saw in Chapter 4, that this disagreement between the experimental scientist and the rationalist is not a new phenomenon; it goes back to the Taoist Sages of the Waring States Period of Ancient China. We also saw that it was nature mysticism that allowed the experimental sciences to overcome the opposition of the dogmatic theologians, theoreticians, and philosophers.

However, in our modern world, Nature and a study of Nature is often invoked by those opposed to science. Perhaps it is in the world of artistic inspiration that we find some of the most thoughtful, and therefore useful (perhaps even persuasive) arguments for anti-science. The poet and mystic, William Blake (1757–1827) reacted particularly strongly against the ideas of Isaac Newton in his paintings and writings, and is seen as being perhaps the earliest, and almost certainly the most prominent and enduring, example of what is seen by historians as the aesthetic, or romantic response against science. In Blake's 1795 poem *Auguries of Innocence*, the poet describes that beautiful exemplar of Nature, the robin redbreast imprisoned by the materialistic cage of Newtonian mathematics and philosophy. In Blake's painting (1795) of Newton (Figure 16.1), Newton is depicted as a misguided hero whose attention was only directed to the drawing of sterile, geometrical patterns on the ground, while the beauty of Nature was all around him; as Blake put it, "*May God us keep / From single vision and Newton's sleep!*" Blake thought that Newton, Bacon, and Locke with their emphasis on mechanistic reasoning were nothing more than "*the three great teachers of atheism, or Satan's Doctrine.*" Blake's painting of Newton, progresses from exuberance and color on the left-side, to sterility and darkness on the right-side. In Blake's view, Newton brings not light, but night. In a poem, W.H. Auden summarizes Blake's anti-scientific views by saying that he "[broke] *off relations in a curse, with the Newtonian Universe.*" But Newton was a complex, universal personality. As we saw in Chapter 1, Isaac Newton was as much at home in the Kabbalah, and other such metaphysics as he was in classical mechanics, but this was not widely appreciated in Blake's day.

Figure 16.1: William Blake's Newton (1795) demonstrates his opposition to the "single-vision" of scientific materialism. Image from: https://en.wikipedia.org/wiki/William_Blake#/media/File:Newton-WilliamBlake.jpg.

Issues of anti-science are best seen as a consideration in the on-going transition from pre-science or proto-science to present-day science. This is what we spoke about in the continued use of the *I Ching* and Astrology as means of divination, and as limited models of the complexity of the natural world. This same argument is evident in the evolution of alchemy (a mystical and an experimental art) into purely functional experimental chemistry. Many disciplines that pre-date the widespread adoption and acceptance of the quantitative scientific method (early 18th century in Europe), such as mathematics and astronomy, are not seen as anti-science. However, some of the orthodoxies within those disciplines that predate a scientific approach, for example, those orthodoxies repudiated by the discoveries of Galileo are seen as being a product of an antiscientific stance. Of course, an ardent or zealous belief in the central importance, the universality and the unfailing potential of science can be considered as a new religion. A religion in which scientists would be the priestly caste. But this would be a religion based upon reproducible miracles. And if everyone could see and benefit from a miracle, then all would become believers.[43]

The derogatory term "scientism" derives from the study of science, and is a term invented and used by social scientists and philosophers of science to describe the views, beliefs, and behavior of strong supporters of science; those who speak of science triumphalism, the science of the late 19th century. It is commonly used in a pejorative sense, for individuals who seem to be treating science in a manner similar to that used by believers in their particular religion.[44]

Of course, it is often the case that a difference arises between scientists and non-social scientists, because of a difference of perception. Some supporters of anti-science may have presented unreal images of science that threaten the believability of scientific knowledge, or appear to have gone too far in their anti-science deconstructions. The question often lies in how much scientists conform to the standard ideal of communalism, universalism, disinterestedness, originality, and skepticism. Unfortunately, scientists don't always conform; scientists do get passionate about pet theories; they do rely on reputation in judging another scientist's work; they do pursue fame and fortune via research. Thus, they may show inherent biases in their work. Indeed, many scientists are not as rational and logical as legend would have them, but then neither are they as illogical or irrational as some supporters of anti-science might say. We are all human.

A point of contention often presented by supporters of anti-science involves the inappropriate, or inadequate nature of the mathematical models used to model real systems. That these models do not capture the full reality of existence. Scientists would be told that the formulae of mathematical models are artificial constructs, logical figments with no necessary relation to the outside

---

[43] This is the premise behind a great deal of science fiction, particularly, the novels of H.G. Wells, such as *The War in the Air* of 1908 and his writings such as *The Shape of Things to Come* of 1933, which were turned into the 1936 movie, *Things to Come*.

[44] Thomas Henry Huxley (1825–1895) was an English biologist and anthropologist specialising in comparative anatomy. He is known as "Darwin's Bulldog" for his aggressive advocacy of Charles Darwin's theory of evolution, an ardent advocate, who would not have been out of place in a mediaeval search for heretics.

world. That such models always leave out the richest and most important part of human experience: daily life, history, human laws and institutions, the modes of human self-expression. That these models fail to appreciate the subtle complexity of the social world; so a great deal of what is best in society is excluded from the model, which, not surprisingly only generates oversimplifications. A great deal of this criticism is true—our models of the world are limited. But they are limited by the ability of present-day computers to solve the equations that describe the model; that is, the models we use are limited by the present limitations of technology. Today's computers are fast, but the computers of the mid-century will be a lot faster, and so better for solving the types of complex social problems that supporters of anti-science criticise scientists for not solving—just be patient.

This difficulty in communicating the evolving interaction of science and the wider society is as the heart of the problem about the public understanding of science, and the vanquishing of anti-science. For my part, I continue to believe in science, and that science has done incalculably more good than harm to mankind. I have not lost faith in science as part of the highest civilization, and its development as one single epic story for all humanity. It might have been the American taxpayers who paid for the journey of the first men to the Moon half a century ago, but it was all mankind who exalted in the achievement (see Figure 16.2).

16.3 THE LIMITATIONS OF THE ENLIGHTENMENT

Supposedly, the Enlightenment of the 18th century was the moment when the light of reason was focused into the obscure corners and occult recesses of the human mind. Yet, the 19th century saw an extraordinary revival of interest in magic, spirituality, and religion. True, there was also the great synthesis by James Clerk Maxwell of electromagnetism, and the first steps toward the triumphs of 20th-century physics, but why was it that so much enchantment survived the Enlightenment's rational examination? Why was it that the Enlightenment's spirit of "daring to know" failed? This question is at the heart of why it is that man has not been translated into a new Golden Age by the extraordinary achievements of science over the three centuries since science came into its own. The German philosopher, Emmanuel Kant, who was the man who told us to "*dare to know*" also said that man, because of his reason was fated to propose and worry about questions he could never an-swer or dismiss. Kant regarded this fruitless search after mysteries to be an aspect of human reason. An irritating aspect that may be deflected or ignored, but cannot be fully denied.

The Enlightenment's quest for any truth that may have been hidden within the occult sci-ences was not a search for magic *per se*. It was a process of copying scientific investigations that in the fields of natural sciences were yielding new useful, verifiable discoveries in Nature. The natural world was seen to be yielding up her secrets to the empirically minded scientist. Those 18th-century occultists and Neo-Platonists would have considered themselves as men of science. It was just that the sciences they pursued were not limited to chemistry, physics, botany, etc., but also include necromancy, alchemy, and magic. Everything was being studied; the practitioners were daring to know everything, and by so doing gave to the more esoteric and recondite subjects a veneer of respectability. The Book of Nature had no forbidden chapters. The rationality of the Enlightenment collapsed into a myth of the type that rationality was intended to banish. Physics became mixed up with metaphysics, and it was Isaac Newton who had warned us to avoid such a mixing, even though his Neo-Platonic outlook told a very different story.

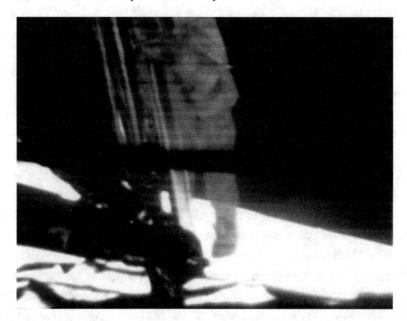

Figure 16.2: Neil Armstrong (1930–2012) became the first human to step onto the surface of the Moon (image from: https://en.wikipedia.org/wiki/Moon_landing#/media/File:Apollo_11_first_step.jpg). He was Commander of NASA's Apollo 11 mission, and is here seen descending the ladder of the Apollo Lunar Module to step onto the Lunar surface. During this descent he spoke one of the most celebrated of all phrases; a phrase that still inspires and transcends tedious earth-bound politics. The video of this event, which the author watched live on TV as an impressionable 13-year-old, may be found at: https://commons.wikimedia.org/wiki/File:Apollo_11_Landing_-_first_steps_on_the_moon.ogv.

It was in the last century that we witnessed the most significant re-appraisal of our well-established way of looking at Nature. At the dawn of the 20th century, we began to finally get to grips with the question of the nature of light, which is just as well as it is via light that we perceive the world around us; and so try to begin to disentangle our perceived sensations from our real and imaginary fears. Physics had shown that light could be thought of as a wave when it propagated freely, yet could also be analyzed successfully if it were considered a stream of tiny particles. Which was true? Was light continuous, or was it particulate? The same could also be said for electrons, so what was the relationship between electrons and light?

Such debates led to the creation of quantum theory, and then the rationalization of quantum theory by Paul Dirac (1902–84) and Werner Heisenberg (1901–76); involving the interpretation of radiation-matter interactions in terms of an uncertainty as to the precise values of the velocity and position of the quantum particles; the problem of complementarity. This rationalization means, that at the quantum level of Nature, measuring something will interfere with the actual validity of the measurement. Niels Bohr (1885–1962), one of the founders of quantum mechanics, thought that the apparatus in which an experiment was performed should be described in the mathematical equations defining what was being measured. Thus, scientific objectivity disappeared. Albert Einstein, the other founder of early quantum mechanics was not happy with Bohr's ideas, and he could never bring himself to abandon belief in the reality of an external world controlled by causality; a world that could be investigated in an objective manner by science. For the purely classical, pre-quantum view of the Universe, we must go back to Pierre-Simon Laplace (1749–1827), who thought in purely classical terms, and who said that from the known laws of mechanics and from a full knowledge of the present state of the Universe, every future state could, in principle, be predicted. But that conceit of Laplace was the view of the European Enlightenment; it was based on their devotion to the mechanical universe of Isaac Newton.

The European Enlightenment did not long survive the Revolutionary Wars of Napoléon Bonaparte. The restoration of the French monarch in 1815 was part of the rapid return to the influence of religion in the political affairs of Europe. For many people, the European Enlightenment was supposed to have done away with ideas of magic and the occult, yet it is true to say that Paris at the end of the 19th century was filled with individuals searching for new ways of looking at Nature. Not only did *fin de siècle* Paris have Picasso inventing cubism, but the greatest, and the most famous physicists in the city at this time were the husband and wife team of Pierre (1859–1906) and Marie Curie (1867–1934), who investigated radioactivity. Yet for all the tremendous research carried out by scientists like the Curies, and the supposed triumph of the Enlightenment, Paris in the period 1880–1914 was also the world center for occultism and mysticism. There was the revival of Rosicrucianism, Helena Blavatsky was attempting a synthesis of eastern and western Hermeticism into what she termed Theosophy (God's wisdom), and symbolism was a dominant idea in literature and music. Why this revival of occultism and spirituality in the City of Light and science? Perhaps because the Curies were investigating radioactivity and radioactive decay; that is, the transformation of one chemical element into another element. This was something that science had always said was impossible, yet was now the hottest topic in physics. How could, supposedly indivisible, eternal atoms such as Uranium decay to form other indivisible, eternal atoms such as Radium and Polonium; what was going on in that atom of Uranium (see Figure 16.3)?

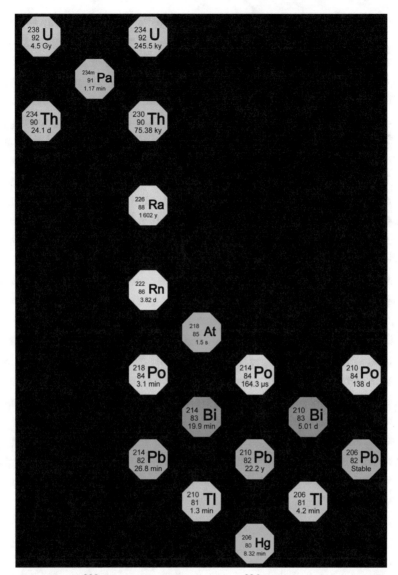

Figure 16.3: Decay chain of ^{238}Uranium, the progenitor of ^{226}Radium. Since the early modern period, Hermeticists and other followers of Hermes Trismegistus had been told they were crazy, and that there was no such thing as the transmutation of the chemical elements. Yet at the end of the 19th century in Paris, the Curies were unravelling the decomposition of a particular isotope of Uranium, and discovering that it transformed into many different chemical elements. Image from: https://en.wikipedia.org/wiki/Radium#/media/File:Decay_chain(4n+2,_Uranium_series).svg.

As far as the 19th-century Hermeticists and alchemists were concerned, the Curies were investigating and seeking an explanation for the transmutation of the chemical elements, something they had always believed in. The Curies were painstakingly showing that many, supposedly eternal, indestructible atoms, could transmute into atoms of other elements releasing huge amounts of energy in the process. Marie Curie even demonstrated how Radium, or rather the radiation emitted by Radium as it transmuted, could cure cancer. All these strange and magical new discoveries in science were to the occultists a vindication of alchemy, and if alchemy was now seen to be true what other occult, or Hermetic sciences, would also be vindicated? But then, man's passion for the fantastic is such that he is only too ready to suspend belief in the rational and the mundane.

In this intellectual ferment, it was not surprising that Einstein and Picasso began coincidentally exploring notions of space and time. Relativity in its overthrow of absolute space and time teaches us that in thinking about perspective, we cannot simply trust our senses; and the cubism of Picasso destroyed the primacy of perspective in art. Indeed, one could say that cubism is art inspired by a redefinition of space and time; a technique for reducing the artistic form to geometry, of representing three dimensions in two dimensions. In their different ways, both Einstein and Picasso discarded the empiricist view—what you see is what you get—in favor of an intellectualized view of the world. But, of course, this re-intellectualization of our study of Nature was the opposite of what had happened in the Middle Ages when modern science had been born in Europe (see Chapter X?). Then a combination of empiricism and nature mysticism had overturned the Aristotelian-Scholasticism of the Middle Ages. At the beginning of the last century, some of the leading figures in the arts were returning to a cerebral, scholastic interpretation of Nature. Proclaiming that thinking, not seeing leads us closer to the truth. Yet the purpose of science is not to provide the most economical representation of the facts, and the purpose of art is not to provide the most accurate representation of what we can see—why compete with photography? The purpose of both science and art is to discover the reality that lies hidden behind the appearances. After all, what is today considered to be magic and science fiction could well become scientific dogma in another half century.

Author Biography

Jeffrey H. Williams was born in Swansea, UK, in 1956. He attended the University College of Wales, Aberystwyth and Cambridge University, being awarded a Ph.D. in chemical physics from the University of Cambridge in 1981. Subsequently, his career as a research scientist was in the physical sciences. First, as a research scientist in the universities of Cambridge, Oxford, Harvard, and Illinois, and subsequently as an experimental physicist at the Institute Laue-Langevin, Grenoble, which remains one of the world's leading centers for research involving neutrons, especially, neutron scattering and diffraction. During this research career, the author published more than seventy technical papers and invited review articles in the peer-reviewed literature. However, after much thought, the author chose to leave research in 1992 and moved to the world of science publishing and the communication of science by becoming the European editor for the physical sciences for the AAAS's *Science*.

Subsequently, the author was Assistant Executive Secretary of the International Union of Pure and Applied Chemistry; the agency responsible for the world-wide advancement of chemistry through international collaboration. And most recently, 2003–2008, he was the head of publications and communications at the *Bureau International des Poids et Mesures* (BIPM), Sèvres. The BIPM is charged by the Meter Convention of 1875 with ensuring world-wide uniformity of measurements, and their traceability to the International System of Units (SI). It was during these years at the BIPM that the author became interested in, and familiar with the origin of the Metric System, its subsequent evolution into the SI, and the coming transformation into the Quantum-SI.

Since retiring, the author has devoted myself to writing. In 2014, he published *Defining and Measuring Nature: The Make of All Things* in the IOP Concise Physics series. This publication outlined the coming changes to the definitions of several of the base units of the SI, and the evolution of the SI into the Quantum-SI. In 2015, he published *Order from Force: A Natural History of the Vacuum* in the IOP Concise Physics series. This title looks at intermolecular forces, but also explores how ordered structures, whether they are galaxies or crystalline solids, arise via the application of a force. Then in 2016, he published *Quantifying Measurement: The Tyranny of Number*, again the IOP Concise Physics series. This title is intended to explain the concepts essential in an understanding of the origins of measurement uncertainty. No matter how well an experiment is done, there is always an uncertainty associated with the final result—something that is often forgotten. In 2017, he published *Crystal Engineering: How Molecules Build Solids* in the IOP Concise Physics series. This title looks at how the many millions of molecules, of hugely varying shapes and size can all be packed into a handful of crystal symmetries. Most recently, 2018, the author published *Molecules as Memes*, again in the IOP Concise Physics Series. This title explains how the onetime separate sciences of physics and chemistry became one science, with the advent of quantum mechanics and the acceptance of the existence of molecules.

In addition, retirement has allowed the author to return to the research laboratory and he is again publishing technical papers, this time in the fields of crystal design and structure determination via x-ray diffraction, in particular, the architecture and temperature stability of co-crystals and molecular adducts.

Printed in the United States
by Baker & Taylor Publisher Services